D0916855

Working Virtually:

Challenges of Virtual Teams

Robert Jones

Robert Oyung

Lise Pace

IRM Press

Publisher of innovative scholarly and professional information technology titles in the cyberage

Hershey • London • Melbourne • Singapore

Acquisitions Editor: Renée Davies
Development Editor: Kristin Roth
Senior Managing Editor: Amanda Appicello
Managing Editor: Jennifer Neidig
Copy Editor: Sue VanderHook
Typesetter: Cindy Consonery
Cover Design: Lisa Tosheff
Printed at: Yurchak Printing Inc.

Published in the United States of America by
IRM Press (an imprint of Idea Group Inc.)
701 E. Chocolate Avenue, Suite 200
Hershey PA 17033-1240
Tel: 717-533-8845
Fax: 717-533-8661
E-mail: cust@idea-group.com
Web site: http://www.irm-press.com

and in the United Kingdom by
IRM Press (an imprint of Idea Group Inc.)
3 Henrietta Street
Covent Garden
London WC2E 8LU
Tel: 44 20 7240 0856
Fax: 44 20 7379 3313
Web site: http://www.eurospan.co.uk

Library of Congress Cataloging-in-Publication Data

Jones, Robert, 1955-
Working virtually : challenges of virtual teams / Robert Jones, Robert Oyung, and Lise Pace.
p. cm.
Summary: "This book provides an in-depth, practical perspective on the growing dependence of virtual teams and how to best exploit them"--Provided by publisher.
Includes bibliographical references.
ISBN 1-59140-585-8 (hard cover) -- ISBN 1-59140-551-3 (soft cover) -- ISBN 1-59140-552-1 (ebook)
1. Virtual work teams. 2. Teams in the workplace--Computer networks. 3. Virtual reality in management. I. Oyung, Robert, 1966- II. Pace, Lise, 1964- III. Title.
HD66.J6565 2005
658.4'022--dc22

2005004519

British Cataloguing in Publication Data
A Cataloguing in Publication record for this book is available from the British Library.

All work contributed to this book is new, previously-unpublished material. The views expressed in this book are those of the authors, but not necessarily of the publisher.

Working Virtually:
Challenges of Virtual Teams

Table of Contents

Section III: Virtual Team Models

Foreword

For over a decade, people have been talking enthusiastically about the potential for telecommuting and other forms of electronically enabled remote work. But most people I know today would say it hasn't happened. They still have offices; they still drive in to work; they still have face-to-face meetings.

What they don't seem to realize (even though it's all around them) is that virtual work is everywhere today. Most professionals I know read and send work-related e-mail from home. Many travel with laptops, cell phones, and other devices that let them stay in touch electronically on business trips and even vacations. More and more meetings include people who are present by speakerphone or are held entirely as conference calls. A growing minority of people now work primarily from a home office—sometimes thousands of miles away from their company office.

When I spent a sabbatical year in Barcelona, Spain, two years ago, for instance, I was surprised at how easily I was able to continue participating in several MIT research projects. We had regular weekly meetings that several participants and I usually attended by phone. We had frequent e-mail exchanges and phone calls. Even though there were a few times when I really wished I had been in a meeting in person, overall I felt like my remote participation in the projects was at least 80% as effective as if I had been there in person.

But all these new possibilities have only become economically feasible in the last decade or so. Taken together, they lead to some profound changes in how teams can work together. But very few people realize just how profound these changes really are.

Now—finally—someone has written a book about what's really happening and what it means. In this remarkable book, Robert Jones, Robert Oyung, and Lise Pace describe what it's really like to live and work in a company that is leading the way in the daily use of virtual teams. They shatter some of the myths that keep many organizations from ever seriously trying to take advantage of virtual teams. They give detailed advice about what works and what doesn't work in virtual teams. They describe the advantages and disadvantages of different kinds of virtual teams and tell a person how to recognize when virtual teams are appropriate and when they are not.

Filled with eminently practical suggestions you can use today, this book also conveys a compelling vision of where we are headed. It's hard to imagine an organization today that could not benefit from reading—and taking to heart—the practical wisdom in this book.

Tom Malone

Professor, MIT Sloan School of Management

Preface

A mere 16 years ago, one of the authors of this book moved from Atlanta, Georgia, to Silicon Valley, California, in order to take advantage of a job opportunity with a different organization in the same company. The company paid his moving expenses, bought his house in Atlanta, and paid points and closing costs on a new house in Fremont, California. The move cost the company over $50,000 and uprooted the author from friends and family.

Today, it would be hard to imagine the same thing happening in the author's same company. The company wouldn't want to spend the $50,000 relocation costs, although such things still do happen in certain circumstances, and the author wouldn't even think of moving 3,000 miles just to switch jobs within the same company. The thing that has changed in those 14 years is the rise of virtual teams.

As we pondered the idea of writing this book, we had several (remote) meetings where we discussed: What do we mean by a *virtual team*? Do we mean geographically dispersed teams, where team members live or work in different locations/states/countries? Do we mean teams with telecommuters (a form of geographic dispersion in itself)? Do we mean teams formed horizontally across vertical organizations (project teams, task forces, etc.)? Do we mean teams formed across different companies? We decided that we meant all of those things, and this book will cover all those manifestations of virtual teams, with a special focus on geographic dispersion.

In the 1990s, businesses began experimenting with various virtual team models to solve specific business problems. These problems included getting the right person for the job regardless of location, getting sales reps out of the office and in front of customers, forming teams that cross national boundaries, and downsizing office space. As we move into the 21st century, additional business problems with virtual teams are being addressed, including the need to cross company boundaries, the increased avoidance of travel (for economic and security reasons, among others), and an increased interest in taking advantage of the inputs from people with diverse perspectives and experiences. Have virtual teams helped to solve those business problems? Are there ways to make virtual teams more effective to better solve those (and yet undefined) business problems? We'll examine those (and other) questions in this book from the perspective of everyone involved in them—participants, program managers, and managers.

Oftentimes, virtual teams are formed to foster cross-organizational interaction in companies that are traditionally organized in vertical "silos." The first virtual team that two of the authors of this book worked on was in the mid-1990s, when they tried to form a sense of community among various decentralized (and globally dispersed) IT organizations that one might call the "congress" approach. In more recent times, our company has gone through a "reinvention," which has both centralized a formally decentralized company and also put much more focus on cross-organizational cooperation and synergy. While suddenly cross-organization teams have become the philosophical norm within our company, there still have been many cultural barriers to overcome along the way.

Within our own organization, we've recently spearheaded a project that allowed engineers in different vertical "silos" to "bid" on projects on which they wished to work within the overall organization. We'll report the results of that particular project in this book.

Regarding geographically dispersed teams, many companies followed a similar progression in the creation of virtual teams. Many started with teams comprised of multiple co-located employees with only one or two individuals working in alternate locations (remote sites, telecommuting, etc.). This was often done either to reward a favored employee by letting him or her work at home (probably the wrong reason to have telecommuting) or when a need arose for a specific skill set that could not be filled locally. With this model, the one or two remote employees often were made to feel like the "odd man or woman out," and team activity centered around the central "hive" where the bulk of the team members worked. The remote team members would "fly in" to the central hive from time to time for face-to-face meetings.

In time, companies started to experiment with models where multiple team members were spread across different sites or even different countries. There was usually a central hive to which team members were expected to travel from time to time, but sometimes the meeting location for the face-to-face meetings would rotate between the sites where the larger concentrations of team members resided. With everyone "remote," team dynamics changed dramatically, and meeting methodologies became less hive-centered (with the attendant "odd-man-or-woman-out" syndrome mentioned previously) and more focused on the efficient use of electronic collaboration tools. (It should be mentioned that as recently as a year ago, our own organization was still having meetings where the hive core would meet in a conference room with especially poor speakerphones used for the "remote dial-in users," who actually were in the majority numerically.)

As teams became more comfortable working remotely, their use of tools such as audio conferences, teamspaces, e-mail, and data conferencing became a norm rather than an exception. However, it was still common for companies to fly people to occasional face-to-face internal meetings (e.g., kick-offs, check points, end games), because, of course, "face-to-face is best." (We even have seen recent industry analysts' articles saying that it's impossible to have effective geographically dispersed teams without occasional face-to-face meetings). Then the "dot-bomb" collapse occurred, corporate earnings plunged, and suddenly the expense of traveling for internal meetings seemed extravagant when companies were laying off thousands of employees. Companies turned more and more to electronic meeting solutions. This trend was fueled by security and safety concerns after the September 11, 2001, terrorist attack on the United States, as well as the SARS epidemic in 2003.

We believe that as we move into the future, various driving forces such as the economy, world political tension, work/life balance, personal preferences, and a generational change will lead companies to use virtual teams as a norm, and discover that the virtual experience may be preferable to meeting face-to-face. Instead of face-to-face meetings as a norm, they will be viewed as a "nice-to-have" feature in good economic times, but not a requirement. (Don't get us wrong—we don't mind meeting face-to-face from a philosophical viewpoint; we just don't view it as a requirement for getting the job done.)

While most of the experiences that we'll recap in this book regarding virtual teams is focused on large enterprises, we believe that many of the precepts and best practices described can be applied to smaller companies or even to community organizations and church groups. One company that has less than 25 employees is organized almost completely virtually, both organizationally and in terms of geographic dispersion.

The company, which is in the unglamorous but profitable business of testing commercial air conditioning units, would never describe itself as especially progressive or in the forefront of an organizational revolution, yet only three of the employees are centrally located, with the rest spread throughout the states where the company does business (i.e., technicians in different geographic regions). Most communication with the small central office is by cell phone, the Web, e-mail, and internal face-to-face meetings that occur once a year or less. Interestingly, the company has been toying with the idea of eliminating the physical central office entirely (it serves no customer function) and having the central office functions provided remotely by telecommuting. Now *that* is a virtual organization!

We believe that the virtual team model will expand in the future, as the need for speed and rapid information exchange overcomes traditionally vertical corporate models. And we believe that geographic dispersion will continue to be a common model in the future, and that people will be much more comfortable working remotely, as the younger generation raised with e-mail, chat rooms, and the Internet integrate into the work force (and as technology continues to improve).

In this book, we hope to share our combined 25+ years of experience working on virtual teams as well as present information and insights from other companies involved in virtual teams with whom we've worked. We'll discuss how we've worked to make virtual meetings better than face-to-face and how virtual teams can help solve real 21st-century business problems. We'll also discuss how geographically-dispersed teams can work well together—perhaps even better than co-located teams. Note that this book is not a beginner's guide, and it assumes some working knowledge of the virtual team concept.

Robert Jones

Robert Oyung

Lise Pace

March 2005

Note: This book was entirely written by three authors in three separate locations with no face-to-face meetings.

Acknowledgments

We'd like to thank all those who agreed to be interviewed for this book, including:

From Eddy Current Specialists:
Ken Eisenhauer and Debra Kasson

From Hewlett-Packard:
Howard Bain, David Brehm, Brandt Faatz, Conor Gavin, Joe Gerardi, Lyle Harp, Brian Jemes, Takehiko Kato, Geoff Markley, Loyal Mealer, Sonja Poling, Greg Todd, and Carol Wolf

Thanks to Tom Malone and Rob Laubacher from MIT who share our passion for virtual work.

Robert Oyung, Robert C. Jones, and Lise S. Pace

Special thanks to Angela, Robin, and Shelby for their inspiration and wonderful ideas.

Robert Oyung

Special thanks to Debra for her encouragement and inspiration during this project.

Robert C. Jones

Special thanks to Marlis, Byron, Jack, and Callahan for their support and patience.

Lise S. Pace

Reader's Guide to *Working Virtually: Challenges of Virtual Teams*

Before we go into too much detail about virtual teams, we should think about why we even have them in the first place and what can be accomplished through virtual teams.

An Introduction to Teamwork and Communications

From an overall perspective, virtual teams are just another model for getting work done. Virtual teams and teams in general are characterized by the fact that each member of the team is dependent upon one or more other members in order to accomplish the overall goal. In contrast, a group is just a collection of people who happen to be working together, but they are not dependent upon each other.

For example, one can hire a group of photographers to take pictures at a wedding. They happen to be at the same place doing the same thing, but they work totally independent of each other. The quality and quantity of a particular photographer's work is completely unrelated to the experience and skill set that the other photographers bring to the wedding.

On the other hand, when a caterer comes to serve lunch at a wedding, the caterer brings a team. The team will cook, set up, and serve. Each member of the team is dependent upon the other to be successful. If the cook doesn't show up, there's nothing to serve. If the set-up team doesn't show up, the cook can't cook.

Thus, situations where each team member's work is interdependent and required for success, the virtual team model can be used to make the team more efficient, more effective, or both.

The single most critical component that makes teamwork possible is effective communication. Again, if you are working in an independent group of people, you don't need to communicate with each other to be successful. But if you are working in an interdependent team, communication is critical—you need to know what the other person is going to do, when they will do it, and how you will work together.

As early as 900 B.C., the first postal service for government use was developed in China. Later, smoke signals, drumbeats, carrier pigeons, and semaphore flag signaling were used as communication tools. These were the tools that enabled long-distance teamwork and, to some extent, the first virtual teams.

The quality and speed of communications drive the effectiveness and efficiency of the team.

Imagine two teams, each on opposite sides of a wide river, trying to communicate with each other through smoke signals on a windy day in order to decide whether they will meet at the head or at the mouth of the river. It takes quite a bit of time to light a fire, which reduces the efficiency of the team. And the wind blowing the smoke makes the quality of the communication low, increasing the chance that the signal will be misread and reducing the effectiveness of the team.

Efficiency and Effectiveness: Divide and Conquer

Divide and conquer is a technique for breaking down a problem into smaller parts, working on the smaller parts, then combining the smaller results into an overall solution.

Let's say you want to build and sell a table. You could easily do everything yourself, if you're building a simple table. You can go out and buy the wood and the stain. Then you design the table, cut the wood, assemble the table, stain it, put an ad in the paper, wait for phone calls, and work with customers.

That's fine if you have quite a bit of time or you only want to sell a few tables a year. But what happens if you want to sell more? What happens if your customers don't want tables made with wood that is local to your shop, but they want exotic foreign wood from across the world?

In that case, it is more efficient and effective to divide and conquer. Perhaps you act as the overall coordinator and designer, and have a team that does the rest. You can have someone else buy the wood from both local and foreign suppliers. Someone in the shop can cut the wood and assemble the table. Someone else can stain it, and another person can advertise and sell the table. You've added quite a bit of complexity to building the table, but now you can handle more requests, and you don't have to do everything yourself. As mentioned previously, communication between team members is critical to ensure success.

Consider another example. Let's say you have a complex question that you're trying to answer. One way to get an answer is to look around the room and ask everyone you see.

That approach looks something like this:

You

That's somewhat of a divide-and-conquer approach. But what if you were to look beyond the resources that happen to be at your current location? You could leverage all the resources you have and all their resources, even though they are not in the room with you. You have more people looking at the problem, more approaches to solving the problem, and, in the end, a more effective solution that was arrived at more efficiently. And with everyone working together towards a common goal, you now have a virtual team. That approach looks like this:

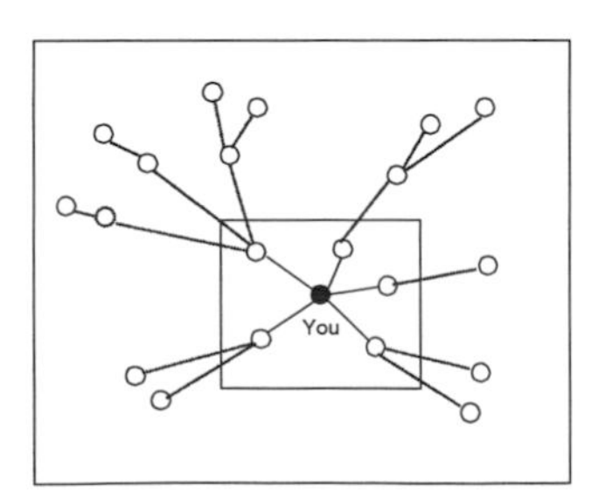

Skill-Sets for Successful Distributed Teams

In a divide-and-conquer approach, teamwork and communication are essential. Making sure you and your team have the right skills for this type of problem solving will ensure that you take full advantage of this technique.

Imagine you are a film actor or actress in 1927. You've spent your entire career developing dramatic facial expressions and exaggerated body motions, because all the films that you've been in have been silent, and that was the only way to get the emotion across to the audience. Then, later in the year, a movie called "The Jazz Singer" is made. It's the first movie with audible dialog. Soon after, silent movies are replaced by movies with sound, music, and dialog. You find that good acting is no longer just about facial expressions and body motions. You now need to learn how to speak with emotion, enunciate your lines clearly, and perhaps even learn to sing. But you also still need to remember the basics of acting that you learned when movies were silent.

Making the transition from working alone to working in a team, and, more importantly, working in a physically distributed virtual team, requires a similar need to acquire new skill sets and continue emphasizing basic ones. Since you will be working with people who are not in the same place as you are and, in many cases, people you do not know personally, communication and trust are the major basic skills that you will need.

Earlier, we discussed the importance of clear and timely communication. Let's look at the importance of trust when working in a distributed team. Without a sense of trust, you are not sure the results you receive from team members are correct, and the team members are not sure why they should be working on the problem in the first place, much less what they get out of helping you with your problem. Establishing this trust and agreement on a common goal will ensure that the team is successful and the result effective.

Throughout the rest of the book, we'll discuss the best practices for gaining this trust and techniques for managing and working in a virtual team, but now you should understand how virtual teams can be used in a divide-and-conquer approach to solving problems and the importance of communication and trust in ensuring virtual teams are successful.

Virtual teams can take many forms. They can include geographically dispersed teams, where team members work in different locations (states, countries, etc.), teams with telecommuters who work away from a central site, teams formed across organizations to produce specific deliverables (project teams, task forces, etc.) or teams formed across different companies.

Virtual Teams: Defined

For the purposes of this book, we define a virtual team as any team with members that are geographically distributed across more than one location. Virtual team members cannot frequently meet face-to-face (i.e., all gathering together in the same meeting room) and must rely on technological tools to facilitate team interactions. Here's how we break down the components of the term *virtual team*.

- **Virtual** - Term used to describe any team with graphically distributed team members who are unable to interact face-to-face on a frequent basis (daily, monthly, quarterly, etc.)
- **Team** - Any group of individuals working together to achieve a business result. For the purposes of our discussion, there are several forms of teams, including the following:

  Project teams - Teams of individuals brought together to produce a specific deliverable (e.g., project teams, task forces, etc.). Team membership can be across organizations and can even include individuals from other companies (e.g., consultants, outsource partners, business partners, etc.). Project teams often are disbanded, once the team's specific deliverable is completed; however, we have also encountered virtual project teams that produced ongoing results over several years.

  For example, a project team for the "Widget Company" might include an engineer from the development department, who works in Germany; a specialist from the marketing organization, who works in Australia; and an engineer from the online support team, who works in France. Management might bring this cross-functional group together in a project team to develop improved marketing approaches for a specific product. Once the marketing plan is assembled, presented to management, and passed on to the appropriate teams for implementation, the project team will disband.

Direct Teams - An individual's direct team is comprised of a manager and his or her direct reports. Members of a direct team also can be geographically distributed. Using the previous example, the development department could have members working in Germany, Australia, and the United States.

An Entire Organization - An entire organization with multiple levels of management and team reporting also can be geographically distributed across multiple locations. In this case, the entire organization would be described as working in a virtual environment.

We will illustrate this with an organizational chart for the Widget Company from our previous discussion.

The Marketing Team, shown in the chart with the solid line box, is an example of a direct team that is geographically distributed, with members in the United States, Germany, and Australia. The virtual project team discussed in the previous example is highlighted with the dashed line. The project team includes Sue (in Australia), Hugo (in Germany), and Beau (in France). The entire Widget Company is a virtual organization, as employees across many teams are located in many different locations. Note that at first glance it may appear that the online support team is a co-located team, with all team members in France. However, even this team could be a virtual team, as members could be working from multiple locations in France.

The Widget Company

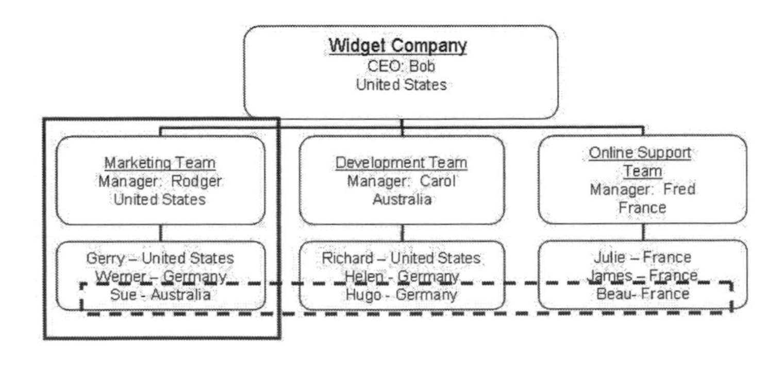

The Format for the Book

A quick scan of the Table of Contents shows that we have divided our content into five major sections. To provide the reader with a framework for our discussion, following are the main elements of each section of the book.

- In **Section I**, we will present what we believe are the eight myths of virtual teams. We will discuss each myth in detail, highlighting why each one has the potential to develop within virtual team environments and how we think their impact can be eliminated (or at least diminished). We'll also present some of the unique challenges that telecommuters face when working on virtual teams and discuss the importance of establishing trust among virtual team members.
- In **Section II**, we will highlight the business benefits of virtual teams and present a case study of a highly successful virtual team (the PC COE team) within Hewlett-Packard.
- In **Section III**, we will discuss the progression of virtual teams within the corporate environment. We will start with what we are defining as a "mostly co-located" model and progress to the "mostly virtual model" for teams. We will also discuss the challenges and needs of virtual teams, the skills necessary for success on virtual teams, and some of the methodologies that virtual teams can use to interact and conduct business. This discussion will include several case studies on creative ways that we've seen virtual teams accomplish typical team goals and activities as well as a chapter on managing a virtual team. We will also introduce what we are labeling a "virtual team maturity curve," designed to help individuals or organizations determine how receptive they are to virtual work as well as presenting an international view of virtual teams.
- In **Section IV**, we will discuss best practices for virtual teams. We will talk about the tools available for use by virtual teams as well as some of the techniques we've seen employed for team building within virtual teams. We'll include a brief discussion of situations where virtual meetings might be preferable to face-to-face interactions. We'll also present some of the arguments against virtual teams.
- In **Section V**, we will look ahead to the future, highlighting how the use of virtual teams within corporations may evolve in the future as well as how future generations may influence the use of virtual teams. We will also

present a narrative that highlights how virtual teams may meet in the 2010 timeframe.

While this book was not written as a beginner's guide to virtual teams and virtual work, if you are new to the area, you may want to scan the Glossary and Tools Appendix to read more about the terms we will be using and the technology tools (i.e., audio-conferencing services, meeting management software, instant messaging, etc.) that we will be discussing throughout the text.

Section I

Virtual Team Myths

Chapter I

Eight Myths of Virtual Teams

Geographically dispersed virtual teams more and more are becoming the norm in many companies. However, not all companies have embraced the idea willingly or gracefully. In Section I of this book, we examine what we call the "eight myths of virtual teams." Note that we don't promise that these issues don't *exist*, rather that in a well run virtual team environment, they *shouldn't* exist.

Myth #1: "It's Always Better to Meet Face-to-Face" (e.g., Project Launches, New Teams, etc.)

Even the most enlightened virtual teammanagers will repeat the mantra, "it's always better to meet face-to-face," and will often treat virtual meetings, education, and team interactions as if they were somehow second best to the wonders and joys of meeting face-to-face. As it turns out, both face-to-face meetings/training and virtual meetings/training have their pluses and minuses—and it isn't clear in today's environment that one has more pluses than the other. Examples of where virtual meetings/training can be superior to face-to-face include cost savings, travel avoidance, flexibility, rush-hour avoidance, and not judging someone on how they look.

Myth #2: "If It's Really Important, You *Must* Do It Face-to-Face" (e.g., Negotiations, Personnel Issues)

As an add-on to the "it's always better face-to-face" philosophy is the viewpoint that "if it's really important, you *must* do it face-to-face." Again, this presupposes that face-to-face is always superior to virtual, especially for things like performance evaluations, interviews, hiring, and negotiations. As it turns out, two of the three authors of this book have never met their manager face-to-face, yet we've both had multiple virtual performance evaluations, with no feeling on either of our parts that we weren't getting proper attention. One of the organizations that we worked for in the past conducted a series of 45 remote interviews in 2001 and hired several applicants—all without meeting the applicants personally. As far as the negotiation angle, while there is some evidence that *contentious* negotiations are best done in person, there is equal evidence that in situations where there is already some trust built up between parties, negotiations can proceed virtually quite successfully.

Myth #3: "Technology Will Solve All Problems" (i.e., We Just Need 3-D Virtual Reality)

As much as we support virtual teams, virtual meetings, virtual education, and so forth, there is no silver-bullet technology that will make it all successful. One of the most successful virtual teams we ever worked on—the HP PC Common Operating Environment team—operated for the first several years of its existence with nothing more than a weekly phone conference, some file shares, and e-mail.

Although we'd *like* to see high definition, 3-D holographic images for our next staff meeting, we'll carry on with what we have until that technology is available.

Myth #4: "You Can't Climb the Corporate Ladder Unless You're Physically There" (i.e., You Need to Sit Near the Boss and Be Seen)

This is an issue that could potentially affect both telecommuters and virtual team members located in company offices (just not the "right" office). It can also be stated as "out of sight, out of mind." While this issue can be an especially tough nut to crack, we have observed over the last several years that with virtual teams becoming the norm in many companies, it is becoming less important to be seen in the office. However, sitting next to a corporate VP is hard to beat for exposure, unless you know how to manage your own "Joe Employee, Inc." marketing, communications, and project management.

Myth #5: "Virtual Communities Are Ineffective" (i.e., Can't Simulate Water Cooler Conversations)

We're all for water cooler conversations—especially if they're about baseball! But managers that still view virtual communities as ineffective because they can't simulate water cooler conversations probably haven't used instant messaging effectively. We keep in daily contact with our geographically dispersed teams through instant messaging, which provides both a real-time communications vehicle and valuable presence information.

Myth #6: "There is a Center of the Universe and Everything Must Revolve Around It" (i.e., People Should Be Willing to Move If They Want to Work For Us)

A mark of doing business in a large corporation from the 1950s through the 1990s was forcing people to move if they wanted to join your company or rise in the ranks. While this had the advantage of getting everyone in one place, it was inefficient because of the expense of the move, the organizational changes that could negate the purpose of the move in a short time, and the potential undesirability of the move from the point of view of the employee and his or her family. In time, as the locales of many corporate headquarters became increasingly expensive places to live (San Francisco, Los Angeles, New York City, Washington, D.C., etc.), companies (because of moving expenses and inability to get the right person for the job) and employees (because of a negative impact on families, especially children) started looking for other options—and virtual teams fit the bill nicely.

Myth #7: "We Need to Focus on Helping Those Few People Who Will be Remote" (Distributed Members Will Be the Exception)

Do you want to destroy your chances of having a successful virtual team from the inception? Here is a quick and easy way to do it—treat your remote members as exceptions. The classic example of this sort of thinking is continuing to have co-located team members meet in a conference room for staff meetings and having remote members dial-in. To compound the problem, the conference room typically has the cheapest, oldest, and poorest quality speaker phone since Alexander Graham Bell invented the telephone.

Our view—if at least 20% of your team is remote, you have a virtual team, and all non-face-to-face meetings should be held virtually.

Myth #8: "Virtual Teams Are About Working At Home" (i.e., It's Only About Telecommuting)

Sometimes people believe that virtual teams = telecommuting, and telecommuting = virtual teams. While the latter is often the case, the former needn't be. In our current team configuration of 25 people, several people are telecommuters, but the majority work in HP offices—seven different ones. One can have a virtual team without *anyone* telecommuting—although telecommuting seems to be an especially reasonable option, if there is no one from your team located in your local office.

In this book, we'll examine the challenges—and advantages—of virtual teaming. And we'll help you to transform the eight items above—which may be realities at your company today—into myths.

Chapter II

Look Them Straight Between the Eyes

When you first meet someone, you start noticing things. You notice what the person is wearing, the color of their hair, their height, the build of their body, what their voice sounds like, and maybe some sort of pleasant or not so pleasant odor. You notice the expression on their face. They may look happy, nervous, serious, angry, or perhaps inquisitive. You begin to form opinions and perceptions about that person and maybe even anticipate the type of interaction you're about to have with them. Some people say they can pick up a certain "vibe" when they meet people. But what happens when these physical cues aren't available? What happens when the first interaction is on the phone or via e-mail? Many people would say, "That's not the best way." Of course, virtual teams do not have the luxury of meeting face to face frequently (if at all) and must do without many of these visual cues.

This chapter will discuss the dynamics behind the myths that revolve around the idea that in the business environment, you must be able to "look them straight between the eyes" to be successful.

- Myth #1: It's always better to meet face-to-face.
- Myth #2: If it's really important, you must do it face-to-face.
- Myth #3: Technology will solve all problems.

Myth #1: "It's Always Better to Meet Face-to-Face"

Most people think that meeting face-to-face is always best and that working remotely in a virtual team is a compromise. In the many years that we've been working on and managing virtual teams, the following are the major benefits of face-to-face meetings that people talk about.

- You just can't build a team virtually—you need to meet people face-to-face.
- You can get more done meeting face-to-face, locked in a room, and just cranking it out.
- Face-to-face meetings provide opportunities for creating meetings to discuss subtopics.
- The face-to-face meeting was good, but the best part was the opportunity to talk to people during the breaks and network with people they otherwise might not see.

Sometimes, it is not feasible or overly inefficient to function in a virtual team environment. For example, it would be difficult and ineffective for a hardware support engineer to repair a piece of complex equipment remotely using the customer as part of the virtual team. Likewise, a doctor performing surgery typically would not be in a situation where working virtually would be effective. However, with improved technology and processes, even these tasks someday may be done routinely in a virtual environment. As a matter of fact, in Australia, a liver specialist assisted surgeons in New Zealand during an operation by guiding them through the procedure based on what he saw through cameras that he controlled remotely.

We're not saying that you should eliminate face-to-face interactions all together, but you need to determine which method is most appropriate for what you want to get out of the interaction. In some situations, face-to-face is better; on other situations, virtual is better.

Face-to-face	Virtual
• Personal interaction is critical • Discussion includes complex media (more than just pitching slides) • Big difference in time zone	• Requires quick startup time (need to meet soon) • Topic primarily discussion, presentation, or sharing of electronic material • Topic can be discussed in one- to two-hour time slots.

Some Approaches to Minimize the Impact

Here are some approaches that we've found helpful in minimizing the impact of the "It's always better to meet face-to-face" myth.

- Instead of using the event as an opportunity or excuse to get together with someone, get into the habit of ongoing discussions and conversations rather than setting up formal face-to-face meetings.
- Create communities by organizing events that give people an opportunity to network—special interest groups, communities of practice.

Sometimes Virtual is Better

Here are some examples of situations where our experience has shown that virtual can be better.

- Virtual team members often can avoid preconceptions based on physical features. Maybe someone looks like they can't be trusted. Maybe their physical appearance conveys a sense of their reliability, dependability, or attention to detail. Research stresses that the majority of communication is conveyed by nonverbal cues. This can be distracting, in the same way that magicians use hand motions to distract you when they're performing a trick.
- Quick ad hoc meetings are easy when they are virtual, either through instant messaging or quick phone calls. Even fast e-mail conversations threads have been known to replace meetings.

Most people probably would agree that typical meetings could be done virtually, but most people agree that there are some situations where face-to-face is essential. For more detailed discussions in this area, see Chapter 21.

Applying Virtual Meeting Techniques to Face-to-Face Meetings

Many virtual teams have discovered that virtual meeting best practices can be applied to face-to-face meetings. Some of these techniques might include the following:

- Clear meeting objectives and agenda (a must for virtual meetings; sometimes considered optional for face-to-face meetings)
- Electronic voting tools (we've used them heavily for things like prioritization sessions)
- Interactive whiteboards displayed on your personal laptop instead of (or in addition to) being displayed as a central projected image
- In large face-to-face meetings, we've even made use of electronic "hands-up" tools, which allow automatic cuing of comments to insure "first-come-first-served" and to allow anyone a forum to speak (not just the loudest or most aggressive)

Myth #2: "If It's Really Important, You Must Do It Face-to-Face"

You're in San Francisco and it's 6:05A.M. You jump out of bed quickly because you're already five minutes late for your audio conference with London. You're half asleep, not because it's early, but because you were on the phone all night working with Singapore during their day preparing for your 6:00 A.M. meeting.

You think to yourself, "Back in the good old days when my team was all local, I wouldn't need to suffer like this." But at the same time, you also would not have been able to have your project completed so quickly by working with experts from around the world instead of trying to do everything in your local team.

"At least," you think to yourself, "we were able to meet face-to-face for our project launch and have a couple of planning meetings in London together. It did take a lot of work to organize those meetings, and we had to push them out a month to accommodate everyone's schedule, but it was worth it."

Was it? Sometimes, a face-to-face meeting is costly and time consuming and unnecessary. In other situations it is very important.

Even with the best technology and the most sophisticated techniques for working successfully in virtual teams, it is still essential that some things be done face-to-face. Most of this is driven by social etiquette.

Should always try to be face-to-face	• Firing an employee • Dealing with a sensitive personnel issue • Hiring an employee
Useful to be face-to-face but not necessary	• Giving bad news • Giving difficult performance feedback • Interviewing for a new job • Discussing a controversial/difficult subject • Meeting with a brand new team
Almost never needs to be face-to-face	• Giving good news • Making a presentation • Meeting with someone new • Meeting with a team member or someone you know • Project launches

Myth #3: "Technology Will Solve All Problems"

Another myth surrounding virtual teams is the idea that technology will solve all of the issues and problems that members of virtual teams encounter as they work together. In other words, virtual teams could be 100% effective if they just had three-dimensional, holographic, virtual reality technologies available to them that would replicate the face-to-face experience to facilitate their interactions. Believers in this myth postulate that, enabled by this type of highly sophisticated technology, life on the virtual team would be ideal, and there would be few other issues that the virtual team would have to address.

This supposition, of course, is not true. Technology alone cannot solve all problems that teams—virtual or otherwise—working together must address.

We've all been on less than effective co-located teams. A scan of any online bookstore will reveal a plethora of titles dedicated to helping co-located teams develop the skills and formulate the shared processes and strategic direction necessary to work effectively. Often, co-located teams conduct many face-to-face meetings but produce very few tangible deliverables due to the failure to address key issues such as fully outlining team goals, documenting project objectives, gaining sponsorship for team activities, addressing cultural and personality differences, developing the team's shared values, and so forth. A virtual team with the latest and greatest technology still can fall victim to neglecting to address these key issues as well.

As we previously mentioned, one of the earliest (and highly successful) virtual teams the authors worked on in the early 1990s relied only on very basic technology tools—audioconferencing, e-mail, and file shares. This was not very sophisticated technology, yet the team spent time addressing the aforementioned issues and was able to be highly successful. (Note that we present a case study of this team in detail in Chapter 8 of the book.)

While better technology can help enable virtual teams to work together more easily, it is not the end-all for success.

Final Thoughts: Some Things Just Can't Be Done Virtually

Although face-to-face is not a requirement for many activities, some things just don't happen when you're virtual. Overhearing conversations is one of those things. There is the basic office gossip that might be entertaining to overhear, but sometimes, you may overhear something that you have knowledge about or information that can answer a question. One of us was in the office the other day, searching for information about how many people we have located at each of our offices. I was asking a co-worker if they knew where I could find the information, and someone sitting close by in another team that I usually don't interact with leaned over and said they had exactly the data I was looking for. That situation would be very difficult to replicate in a virtual environment.

A couple of virtual alternatives to the serendipitous encounter just mentioned are blogging and discussion boards. We discuss blogging in Chapter 18 as a way to publish an electronic journal. Including requests for information in a blog might result in someone who was casually reading your blog to respond with useful data. Posting that same request on a discussion board could be just as effective.

Chapter III

Out of Sight, Out of Mind

Conventional wisdom often cites that individuals cannot advance their career without putting in daily face-to-face time with their managers; thus, members of virtual teams are inevitably less successful at career development than their co-located colleagues. Virtual team members lose out on the informal interactions that typically occur in the lunchroom, in the hallway, at the water cooler, and on the golf course; therefore, they are essentially out of sight, out of mind. Furthermore, conventional wisdom postulates, it is impossible to get the attention of upper management without these interactions, and people believe that if you want to advance, you will move to the companies selected "center of the universe" where the largest concentration of employees reside.

This chapter will discuss the potential drivers behind the following myths that focus on this potential out-of-sight, out-of-mind dynamic, and will document some of the techniques that we've observed that help reduce the potential impact of the reduced face-to-face time that is inevitable with the implementation of virtual teams within an organization.

- Myth #4: You can't climb the corporate ladder unless you're physically there—you need to sit near the boss and be seen.
- Myth #5: Virtual communities are ineffective—they can't simulate water-cooler" conversations.
- Myth #6: There is a center of the universe, and everything must revolve around it.

Myth #4: "You Can't Climb the Corporate Ladder Unless You're Physically There"

Myth #6: "There is a Center of the Universe, and Everything Must Revolve Around It"

First, one key point to remember is that while the out-of-sight, out-of-mind dynamic might exist within teams where there is a very large central presence and only one or two team members working remotely, we believe that the impact diminishes greatly as virtual teams become more and more geographically distributed. In a mostly virtual, geographically distributed organization (which we will discuss in Chapter 10), it is entirely possible that very few (or even no) team members will actually be co-located with their direct managers or supervisors. Upper level management may be similarly distributed as well.

The authors' prior organization (within HP's Information Technology organization) had employees around the world, including Palo Alto, California; Atlanta, Georgia; Fort Collins, Colorado; Grenoble, France; Singapore; the United Kingdom; Corvallis, Oregon; Roseville, California; and Boise, Idaho. Roughly 60% of the organization's employees worked outside the corporate headquarters, with first and second level managers represented in multiple locations. As we write this book, two of the authors work for an organization that has employees in multiple locations including Palo Alto, California; Houston, Texas; Boise, Idaho; Atlanta, Georgia; Dornach, Germany; and the United Kingdom. When working on teams with this level of geographic dispersion, it becomes increasingly hard to calculate where the center of team interactions actually resides.

Corporations are also moving more towards results-oriented human resources processes that are clearly focused on rewarding those who meet goals and objectives, which also assists in ameliorating the out-of-sight, out-of-mind dynamic. Working in a results-based environment with clearly articulated objectives and measures also helps virtual team members. If you can illustrate at the end of every review period (quarterly, semi-annually, annually, etc.) that you have met your business objectives (no matter where you work or who else works at the same location), then you are more likely to be rated well and to advance. These results-oriented reviews that are focused on metrics and deliverables are much less likely to result in scenarios where employees are able to curry favor based on personal relationships with management.

Moreover, the definition of "climbing the corporate ladder" within the corporate environment is changing over time, as well. Management structures in many companies are getting flatter, with intermediate levels of management being reduced or eliminated. In these situations, advancing one's career often means getting a variety of experiences across businesses and organizations rather than moving up multiple layers of management. Systematically trying to move up to the next layer of management within your existing organization no longer may be the most advantageous career move—perhaps moving to another business or area to gain a new skill or experience is the way to go. In this case, relationships built while working on virtual teams that cross organizational boundaries may actually be the stepping stone to advancement. The idea that career advancement depends on a variety of experiences also partners well with the virtual team model—no need to move employees around to join new projects or new organizations to expand their skill sets.

Of course, whether you work on a virtual team or a co-located team, individuals must be diligent in marketing their work and managing their careers. There are a plethora of books on topics such as marketing your skills, selling yourself through your résumé, advancing your career, and so forth, illustrating that the need for individuals to market themselves, their skills, and their accomplishments in the business world is not exclusive to members of virtual teams.

While we aren't implying that advancing one's career in the virtual world does not take effort, visibility is achieved more typically through results and deliverables than through the more casual contact (water cooler, lunchroom, golf course, etc.) often prevalent within co-located teams. While proximity to the corporate suite can give more opportunities for informal interaction with managers and other key decision makers, here are some techniques that we've found helpful in marketing accomplishments within the virtual team environment. Note that most of these items are not unique to virtual teams; however, in the virtual team model, the techniques will be largely electronically based.

Some Approaches Individual Contributors Can Use

- Use status reports, newsletters, Web sites, and other information-sharing mechanisms to your advantage.
- Always be prepared to say a few words marketing your projects and accomplishments.
- Seek out opportunities to meet and interact informally with individuals to help build your network.

Here are some of the ways that we've seen assist individual contributors in addressing the potential out-of-sight, out-of-mind dynamics within virtual teams.

- **Use the status report to your advantage.** This may seem like a no-brainer, but in the virtual team environment, completing routine tasks such as status reports on time while sufficiently highlighting accomplishments becomes particularly crucial. You won't be running into your manager (or your manager's manager) in the hall or elevator to give informal updates during the week, so here's your chance. And don't just include a (boring) blow-by-blow account of everything you did that week—be sure to include relevant metrics and positive feedback you may have received. Summaries of electronic status reports are typically cascaded throughout the organization (and sometimes beyond), and let's face it, your manager wants to look good, too, so it is likely that positive results will be forwarded. These highlights also can be leveraged when you are asked to provide input on your accomplishments for your company's human resources review processes. (Note that good judgment clearly is needed here—if you know you manager wants the boring details, be sure to include those as well.)
- **Provide periodic project updates via e-mail.** Again, this is not that different from the co-located world, but for lengthy projects, periodic updates can keep management informed on progress and accomplishments. In the past, when teams were primarily co-located, people sent e-mails on the weekend and during the night to show their manager and team members that they were working hard. With virtual teams working across time zones, showing that you are working hard is no longer the goal. The goal is to ensure that information and status is being shared on a timely basis.

- **Use Web sites and organizational newsletters to your advantage.** Organizational Web sites and newsletters can be used to showcase accomplishments and provide updates. They also can be used as a "pointer" for updates when you are asked about projects. One of the authors has served as a coordinator for an organizational newsletter and was constantly amazed at how difficult it was to extract articles that essentially would be highlighting accomplishments and communicating progress to larger audiences.
- **Always be prepared to say a few words about your projects.** You never know when you might have a chance to opportunistically provide an update on your accomplishments. Perhaps you might be waiting for an audioconference to initiate when a key manager fills the time by asking, "So, what you are you working on these days?" Be ready.
- **Seek out opportunities to meet and interact informally with individuals (managers or otherwise) to build your network.** Opportunities might exist within mentoring programs, community service programs, and so forth.

Some Approaches Virtual Team Managers Can Use

- Utilize the above techniques to market your team's accomplishments.
- Don't let team dynamics revolve around one location, a "center of the universe."
- Know your team's needs as well as the needs of the "chain of command."

Here are some approaches that managers can use, as well.

- **Utilize the above techniques for marketing both yourself and your team's accomplishments** – Each of the items under "individual contributor" could be applied equally to the virtual team manager. Use all available vehicles to market the accomplishments of your team.
- **Don't let team dynamics revolve around a "center of the universe"** – If a concentration of virtual team members does exist in one time zone or location, don't let all staff meetings, project meetings, and so forth end up occurring solely at the convenience of that time zone or location. This can create the illusion that the world revolves around that particular location and that if you aren't there, you aren't a player.

- **Get to know your team members' needs and the needs of the "chain of command"** - Not all employees thrive on regular (i.e., weekly, bi-weekly, etc.) one-on-ones just to give updates on projects. If employees are comfortable doing these types of general updates electronically, that's great. In a virtual team environment, some members will be more focused on other aspects of the working relationship. For example, a key piece of information to communicate to virtual team members is what process they should use when they need to have decisions made. Is it best to contact you through voicemail, e-mail, or instant messaging to set-up an ad hoc meeting or solicit urgent feedback? Many team members will be very comfortable once they know that you are available and know how to engage participation when needed. Also, make sure you know how your manager (and your manager's manager and their manager, etc.) likes to receive data and updates. Encourage your team to provide information in these formats that are customized for the particular audience being addressed.

At this point, the reader may be saying, boy that sounds like a lot of extra work. Perhaps it is, but as a virtual team member, you are receiving personal benefits (i.e., getting to live where you want to live, etc.), and even if you were on a co-located team, you would still have to market yourself to get ahead in the workplace.

Myth #5: "Virtual Communities Are Ineffective"

The myth that virtual communities are ineffective is an interesting one, as well. Perhaps this myth may have been somewhat true 10 to 15 years ago as virtual teams were just emerging due to factors such as the shortage of tools available to enable electronic communities and lack of experience of the participants in working electronically. However, today, we do see examples of successful virtual communities within the Internet world. For example, the open source community operates largely through electronic means and produces deliverables (products) that are utilized by thousands of users worldwide. Usenet groups and chat rooms have also provided mechanisms for groups with a common purpose to exchange ideas, provide technical support, and so forth. Chat rooms around every sort of hobby are also available on the Internet, and participants often form virtual relationships with other participants.

Case Studies of Several Communities Within Hewlett-Packard

Here are a couple of profiles of communities that we've seen formed within Hewlett-Packard.

- **Strategic Forum Community**. This is a community of approximately 100 to 200 members of HP's IT community that gathers for a monthly one-hour audioconference to discuss a single topic of strategic interest within the IT community. The community has been in operation for approximately five years. The audience for the monthly meeting is between 40 and 60 participants, which rotate, depending on the topic being discussed. When the Strategic Forum was initially launched, the only tool used was audioconferencing bridge lines. Any slides used for the discussion were posted in advance on a Web site for participants to download. The meetings were also recorded for off-line listening for interested parties that were unable to attend the actual meeting. For the past several years, the group has also incorporated a Web-based meeting management tool that allows users to join the meeting (and to see who else is in attendance, as well), have the slides automatically advanced, and post questions online.
- **The PC COE Community.** This community (highlighted in Chapter 8) evolved from HP's efforts to develop a desktop management environment in the early 1990s. Members of geographically distributed IT groups as well as business groups participated in the community, providing feedback and input to the core team on product development, and so forth.
- **Mobility, Collaboration, and Workplace Services Information Sharing Community**. This community, which has been in operation for approximately two years, is comprised of members of a development and support organization interested in sharing ideas on new and breakthrough technologies that could be incorporated into service offerings. The community is comprised of approximately 225 members. Depending on

the topic and personal scheduling constraints, anywhere from 45 to 120 people attend the community's one-hour, monthly virtual meetings. Lyle Harp, the community facilitator, notes the community's overall goal.

> Our goal is to generate technical discussions around breakthrough technologies and solutions.... On each call we encourage dialogue, asking people to make comments and ask questions related to the topic.

The community uses a variety of tools to conduct their monthly meetings and to stay in touch between meetings, as Lyle elaborates:

> We use an international audio bridge and HP Virtual Classroom.... HP Virtual Classroom is used for presentation sharing and live demos, voting and polling, posting questions and answers, and visual queues such as a real time list of attendees, "hand-up," etc. (which contributes to a sense of community during the session). We also have a Web site where we post presentations and audio recordings from past sessions and announcements of future sessions.

Lyle offers this learning from facilitating the community:

> Even when there is a deliberate attempt to keep sessions informal, people are less and less likely to speak up as the group gets larger. For this reason, as the leader of the sessions, I attempt to have some questions ready to ask the presenter to kick off Q&A time. This usually encourages others to join in the discussion.

Lyle also notes that when the month's topic is announced via e-mail communication, the community often grows when others working within the area of interest reply with their feedback and comments.

> More than once we've caused some concerns to be raised when we announce a topic, because someone within the company felt that they should own the respective topic for HP. We always assure them that we

are not 'announcing' any product or service, and we usually try to incorporate them in the presentation.... In an informal way, this helps to fill out the picture of who does what across the company.

Within the author's work on virtual teams, one key technology that has emerged in the last several years to aid in the community building process across virtual teams is the advent of instant messaging. Although instant messaging originally emerged in the consumer marketplace as a way for Web users to determine whether "buddies" were online, we've found that instant messaging has been a key enabler of more casual contact among geographically distributed team members. Instant messaging can also provide a sense of presence that is often absent among remote team members (e.g., Is Bob at his desk? I wonder if Mary is on the phone or available to chat?).

Final Thoughts on the Out-of-Sight, Out-of-Mind Myths

Of course, we are not postulating that the myths discussed in this chapter are completely absent from virtual teams within the corporate environment. In many companies and organizations, the time may arrive when one is expected to move to a central office to continue to advance one's career—at least until more businesses fully embrace the virtual organization model that we will discuss in Chapter 22.

Rather, the out-of-sight, out-of-mind dynamic that has the potential to evolve among distributed teams does not have to be a show stopper for career advancement. Virtual team members are not without tools and techniques that can be utilized to market their accomplishments and advance their careers. And, in some cases, the unique and varied experiences that a virtual team member can assemble for their résumé can often make them more marketable than co-located colleagues that may be more limited in their experiences.

Chapter IV

It's the Exception, Not the Norm

Much of the early literature on managing a successful virtual team or on how to successfully implement a telecommuting program made the assumption that most team members were co-located, and the remote user was the exception. The goal for the virtual team, therefore, was to make sure that these one or two remote individuals were able to work within the team on a best-effort basis rather than one encouraging all team members to act virtually. A good example of where this situation might exist is in programs where telecommuting is positioned as a perk, as opposed to a business benefit.

In this chapter, we will discuss Myth #7 and Myth #8.

- Myth #7: We need to focus on helping those few people who will be remote.
- Myth #8: Virtual teams are about working at home (it's only about telecommuting).

Myth #7: "We Need to Focus on Helping Those Few People Who Will Be Remote"

In our experience, the "distributed-members-are-the-exception" viewpoint is fraught with dangers for successful virtual teaming. First, it is very difficult for a single remote member ever to feel truly integrated into a predominately co-located team. This environment is typically characterized by team behaviors (often unintentional) that reinforce the co-located paradigms.

Second, the "distributed-members-are-the-exception" mindset often leads to the awful "meeting-room-with-a-single-dial-in" syndrome, where a group of co-located people gather in a conference room with a single speaker phone (probably the oldest and poorest quality unit in the corporation) and a single dial-in user joins the meeting. Inevitably, the poor dial-in user not only can't identify who is speaking at any given moment, but usually can barely hear, anyway. In our early virtual team experiences, the authors of this book have been on the receiving end of the "single-speaker-phone" treatment—no one likes this experience. Remote dial-in participants lose the visual queues that everyone else in the meeting room enjoys (nodding heads, bored looks, someone leaving the room, etc.) as well as the typical banter that occurs among the face-to-face participants. The remote team members have only a second-class experience here.

Our view is that there is no such thing as a partially co-located, partially remote team. If a single member is remote, you have a geographically distributed virtual team, and it should be managed as such. And if you have a virtual team, most meetings (almost by definition) should be virtual, using dataconferencing, audioconferencing, and the myriad other electronic collaboration tools that support virtual meetings.

One of the most successful ways to avoid the "second-class-citizen effect" for the remote members of a virtual team is for all team members to dial into the same audioconference bridge, rather than co-located team members gathering together in a room and conferencing in the remote members. Everyone on the team needs to get used to the virtual team protocols, such as saying your name before you speak, calling out slide numbers during slide presentations, and so forth. Note that we've been on conference calls with 30 or more people on the line, and some participants still will use only their first name when speaking or asking questions—not a best practice for virtual teams.

Some team members initially may grumble that the effort involved in adopting virtual work methodologies sounds extreme to accommodate one or two individuals. However, many enterprises will discover over time that their virtual teams will become more—not less—geographically dispersed, as one of the key benefits of virtual teaming is getting the right person for the job, regardless of location. So, getting into the habit of working virtually is a good investment for the future.

Myth #8: "Virtual Teams Are About Working at Home"

Another interesting myth regarding virtual teams is the idea that all virtual team members will be telecommuters. In our view, virtual teams are really about working anywhere and anytime, so telecommuters are just one profile for an individual who may be a member of a virtual team. Team members can also work at geographically distributed corporate sites, customer sites, cars, airports, hotel rooms, Internet kiosks, and so forth. Some will work from all these different locations at different times, based on individual schedules.

However, we acknowledge that telecommuters often have a unique set of challenges when participating as virtual team members, so we'll mention here a few of the key ones that we've encountered in our experience.

- Telecommuters are often left to solve their own technical problems, due to the unique nature of their work environment. This might involve troubleshooting day-to-day technical issues or the setting up the complex integration of their own services such as telephone (through a phone company), broadband (perhaps through another vendor), and so forth.
- Telecommuters often have slower bandwidth connections than their in-office colleagues. Although more pervasive availability of high-speed, broadband services is helping to diminishing the impact of connectivity speed, in-office users probably will always have a slight advantage when it comes to networking speed.
- Telecommuters potentially can experience more isolation and a lack of face-to-face interaction with any in-office counterparts or other company employees (even if they do not work directly with them every day). Going

to the company office can provide a sense of corporate community that a home office may lack.

- Telecommuters often have elevated concerns about being out of sight, out of mind. This particularly can be an issue if there is a large concentration of co-located team members in a central office. Again, the impact of this dynamic is reduced as teams become more distributed. Eventually, when virtual teams are taken to the extreme, no one has the advantage—the team manager might work in Houston, Texas, while his or her manager may work in Boise, Idaho, and so forth.
- Telecommuters often must deal with the perception that since they are not located in an official office environment, they are not really working. Many who are not experienced with telecommuting might perceive that typical telecommuters spend the bulk of their time sitting around their house in a robe and bunny slippers, reading the paper and taking extended coffee breaks to run errands. However, as virtual teams become more distributed geographically, it becomes harder to distinguish where, when, and how team members are working. A team member on Eastern time in the United States may have very little working day overlap with a team member working in Asia-Pacific. Everyone must agree to depend on deliverables and work results to determine if people are actually working.

Final Thoughts on "It's the Exception, Not the Norm" Myth

Unfortunately, the "It's the exception, not the norm" myths are still true in some companies, especially in ones that are fairly new to the virtual team and/or telecommuting arena. Everyone on the virtual team must put forth effort to help debunk these myths.

Chapter V

Establishing Trust

Many of the myths and challenges described previously stem from the fact that there is not a strong established trust relationship within a team. If managers feel that their employees can work independently and deliver on their commitments without constant supervision and micromanagement, then people might feel less need to meet face-to-face.

If employees feel that their work is visible, that their peers understand and value their contributions, and that their managers provide support and encouragement during all phases of their projects, then the employees might not feel that they have to be sitting next to their bosses in order to climb the corporate ladder and to be successful. One manager, Joe Gerardi, explained the importance of building a trust relationship.

> Nobody cares what you think until they know you care about them. Teamwork is more than a series of transactions. Teamwork is a series of engagements. It really depends on the depth of the relationship you have with the person. I wouldn't want to fire someone remotely, but if I had to, I could fire Person A over the phone because we have built a strong relationship.

Consider the following: Even if we have very poor communication and no tools to support a team, if we have a strong trust relationship, there might still be a chance for success. However, just having great communication and the best tools, with no strong trust relationship, doesn't guarantee anything.

Trust is the single most important driver for the success of virtual teams.

If you are able to establish trust as a team member, leader, or manager, you will be able to gain consensus, agreement, and the ability to influence.

Consensus and Agreement

In some groups and organizations, the simple fact that a person is in a leadership position gives that person full power to make decisions, set direction, and tell people what they should do. Although team members may follow orders, they potentially may not be in total agreement with the decision, especially if the direction seems risky or ambiguous, or in some way makes them feel uncomfortable. In the short term, this might result in resentment and later disengagement or non-performance. In the long term, team members might become less independent and look to the leader to make every minute decision, or they may just leave. The most effective group is not the one in which the members are hesitantly participating, but rather the one that is fully engaged and contributing at their highest performance.

Let's consider the following situation: You are leading a team of people who want to go river rafting. Rivers are rated from Class 1 to Class 6, based upon the difficulty of passage. A Class 1 river is basically flat water, appropriate for travel by children. A Class 6 river is so difficult that safe passage is doubtful, and chances for serious injury and death are almost certain. If you are trying to decide what class of river to take, then trust, agreement, and consensus are critical in order to make sure you have a fun but safe trip. Gaining agreement and consensus in this situation ensures that everyone has an understanding of the following:

- What is the overall goal of the activity? (To have fun? To be challenged?)
- How much risk is the team willing to take?
- Which skills are necessary to take on this activity, and, more importantly, which skills are missing from the team?

- What roles will each member have, and are they comfortable with those roles?
- What does each member need to do before, during, and after the activity?

If you do not have agreement and consensus on these questions, you may not have fun, or, worse yet, you or someone on the team may be seriously injured.

Gaining consensus and agreement is important in engaging team members. But how do you do that in a virtual team environment?

The Virtual Handshake

In a face-to-face interaction, you have the opportunity to talk to a person, get to know them, see the emotion on their face, and read their body language. In the river rafting situation, it would be easy to look at your team member's face and see if they were excited, nervous, or disinterested. On the phone, you might be able to hear the tone of their voice, but in an e-mail, it might be very difficult to gauge their comfort level.

Here are a few suggestions to ensure that you and your team members have a virtual handshake to confirm an agreement:

- Communicate with the entire team so they hear the same message at the same time.
- Describe clearly the overall objective, plans, and responsibilities in detail.
- Ask if anyone has questions or concerns. If you get no answer, ask specific team members for their opinion or input. Silence could mean that everyone understands, or it could mean that no one understands.
- Address concerns directly. Be encouraging, but don't try to sugar coat anything so that it sounds better than it really is. Be open, honest, and realistic.
- Follow up individually with anyone that might have concerns they are not expressing in front of the entire group.

If you already have a trust relationship with the team member, it should be easy to come to an agreement or to expose the areas of disagreement. If you don't have a trust relationship with a person, it is very important to build that trust with the suggestions provided, especially if you will be interacting with that person again. In some cases, it is helpful to ask someone that the person already trusts to help you obtain that virtual handshake. By leveraging that existing relationship, you may be able to gain the team member's trust more quickly.

Let's see how our river rafting trip could be managed in a virtual team situation. Cindy is organizing the river rafting trip for several friends who live in different cities across the United States. They are all going to meet in Colorado. Although she knows most of the people who will be joining her, some of them have invited their friends, whom she has never met. She understands the importance of building trust in this team, since they are going on a Class 3 river. She is conscious of the need to have a virtual handshake with the team and will use the techniques described earlier. Cindy starts by arranging a phone conference with everyone. The conversation goes something like this:

> **Cindy:** Hi, this is Cindy. I wanted to make sure everyone knew the logistics for the trip and had a chance to ask any questions. Most of you have had some experience river rafting, so the river we will take next week should be pretty fun but still provide us with a bit of a challenge. Remember, this is a Class 3 river, which many of you have taken before, but we still need to be cautious and prepared so that everyone stays safe. Any questions so far?
>
> (*silence*)
>
> **Cindy:** Max, you've been on a Class 3 river several times. Do you have any advice for people?
>
> **Max:** Sure. Make sure that you're all.....
>
> (*later in the discussion*)
>
> **Cindy:** Mark, we haven't heard from you yet. Do you have any questions?
>
> **Mark:** Actually, I've only been rafting a couple of times, and I just wanted to come along for the ride.
>
> **Cindy:** No problem. You can sit next to Max, and the two of you can be partners.
>
> **Mark:** Okay, that sounds good.

Once you have established this virtual handshake, each interaction you have is an opportunity to build trust in order to continue your partnership in the future. This will be the most important skill used to ensure the success of the virtual team.

Section II

Business Case

The first virtual team that one of the authors ever worked on was in the late 1980s, when Hewlett-Packard consolidated the management of its two North American Response Centers. Suddenly, many employees had a manager (or second-level manager) that was 2,500 hundred miles away! Although this was a bit of a shock at the time, it didn't take long to adapt to it. After all, our job was taking phone-in customer service (PICS) calls, and we were so busy that it didn't really matter if our manager was local (Atlanta) or remote (Santa Clara). What did HP get out of it? A more coordinated and customer-focused telephone support organization, with lower cost.

As it turns out, most times that companies move to a virtual team model, it is because of business pressures to do things such as the following:

- Show me the money
  - Eliminate expensive company moves
  - Travel avoidance
  - Real estate savings

- Human Resources advantages
  - Getting the right person for the job, regardless of location
  - Provide better safety and security
  - Provide work/life balance
- Organizational advantages
  - Tying together a decentralized organization
  - Mergers and acquisitions
  - Regional representatives
  - Getting sales representatives out of the office and in front of customers
  - Innovation
  - Other virtual team drivers

Finally, we'll end up this section with a discussion on "why some companies don't get it."

Chapter VI

Show Me the Money

Company Moves

In 1988, this author received a promotion with a caveat—move from Atlanta (low cost of living) to the San Francisco Bay Area (high cost of living). But in the 1980s, that was the sort of thing that employees did to climb up the company ladder. My wife and I discussed the ins and outs for several weeks before I finally accepted the Bay Area job.

To sweeten the deal, HP paid for my move, with inducements such as the following:

- Purchase my house (spacious and inexpensive) in Atlanta
- Pay my moving expenses
- Pay the points and closing on my new Fremont, California house (small and expensive)
- Several thousand dollars for miscellaneous expenses
- Extra time off for the move

At the end of the day, HP spent several tens of thousands of dollars on my move (nationally, the average in 2003 was $42,000). What did the company and I get out of it?

On the positive side:

- HP got four years of quality (I hope) training courses written and delivered by me.
- I rose slightly up the company ladder and received the largest salary increases to date in my (young) HP career. I also now had factory or division experience, which made me more marketable.

On the negative side:

- Within four months after the move, my wife of 14 years decided that she couldn't take the isolation that she felt after the move, and she moved back to Atlanta (we were later divorced).
- Within six months after the move, there was a significant reorganization at my division, and suddenly my job there no longer existed. I was assigned a new job.
- Within three years of my moving to California, I was selected to participate in a workforce reduction program at my division. Eventually, I found work in another HP division in California.

So was it a good deal for Hewlett-Packard to move me to California? Only if my job absolutely could not have been done remotely. Was it a good deal for me? I lost a marriage and gained some good stuff on my résumé—not much of a net gain for me.

So, if the same opportunity had existed today, I doubt very much that HP would have paid to move me to California, regardless of any unique product skills that I had. I could have researched and written courses remotely and delivered them through the Internet, videoconferencing, or audioconferencing.

A 2001 Atlas Van Lines survey showed a rise in the number of employees declining relocation offers—a sign that there is employee backlash against relocation.

> . . . according to a 2001 Atlas Van Lines survey, the number of companies reporting that employees rejected relocation offers increased from 39 percent in 1999 to 50 percent in 2001. Research

indicates that an increasing percentage of older employees are declining relocation offers because they want to avoid the social and economic impacts associated with moving to a new geographic location, particularly the disruptions to family ties that often ensue.

State of the Workforce 2004: United States. Corporate Leadership Council, February 2004. www.corporateleadershipcouncil.com

Of course, companies continue to pay to move high-level executives after mergers, for example. However, as margins tighten in many industries, moving employees around the country (or around the world) will become less enticing either for the employer or the employee.

Company moves are expensive for the employer and disruptive for the employee—and often unnecessary in the modern world of virtual teams and virtual meetings.

Travel Avoidance

Prior to September 11, 2001, and the dot com bust, traveling for internal meetings was considered not just acceptable, but downright necessary. As more and more companies created geographically dispersed teams, industry publications trumpeted the need for face-to-face kickoff meetings, checkpoint meetings, and end-game meetings. Airlines charged a premium for business travel, and much of it was booked at the last moment.

Then, the terrorist attacks on the World Trade Center happened, and the recession deepened in its aftermath. Suddenly, companies didn't have the travel budgets that were common in the dot com boom days. At first, some companies went with travel only for customer-facing meetings. Some went further with travel only for customer-facing *revenue generating* meetings. And an interesting thing happened (or didn't happen). As in the Sherlock Holmes story "Silver Blaze," where the "curious incident of the dog in the night" was that the dog did nothing, the same happened with extreme travel restrictions—there was no discernable negative impact on projects. Teams still formed, products were still created and sold to customers, and IT departments still managed servers and networks. But one thing did happen—millions of dollars were saved on plane tickets, hotel rooms, rental cars, and lost time. It turned out that travel for internal meetings was really a nice-to-have thing—not a necessity.

Travel savings by using virtual meetings can be significant. The table below assumes a meeting between four people for four hours. The face-to-face expenses assume three of the four must travel an average of 1,000 miles.

Average Cost for a Four-Hour Meeting With Four Participants

Expense	**Face-to-Face With Three Traveling**	**Virtual Meeting**
Airfare	3 x $500	
Lodging	3 x $100	
Food	3 x $25	
Parking/mileage	3 x $32	
Car rental/gas	3 x $80	
Lost productivity	3 x $500	
Phone bridge		$48
Web cast		$120*
Total	$1,255.00	$168.00

*This cost could approach zero if a non-subscription service such as NetMeeting were used. However, we have found NetMeeting to be ineffective when more than 15 people are connected.

As companies become more comfortable with the concept of virtual meetings, the need for travel for internal meetings should continue to fall, even if the economy is robust, and security isn't an issue.

Savings on Real Estate

Virtual teaming, especially when combined with telecommuting, can result in lowered real estate costs, as work becomes more and more something you do and not somewhere you go. An article entitled "Valley Firms Ditch Desks to Cut Costs," by Steve Johnson of *Mercury News*, listed some real estate savings in the technology industry:

- Over the next three to five years, Cisco hopes to cut 15 percent to 20 percent from its real estate costs by eliminating individual offices.
- Hewlett-Packard spokeswoman Brigida Bergkamp cited the April closure of the firm's 495,000-square-foot Mayfair campus in Mountain View as an example. The Palo Alto company was able to absorb the 1,000 Mayfair workers into other buildings, she said, partly because it determined about 455 of them didn't need their own offices."
- About 13,000 of Santa Clara-based Sun Microsystems' 35,000 employees lack offices, according to company executives. That has helped Sun trim at least 500,000 square feet of space over the past couple of years, while reducing its annual real estate bill by $71 million.
- Bill Vass, Sun's vice president of information technology, said having fewer offices hasn't just cut building costs. It's also reduced the company's annual electricity bill $2.8 million worldwide, with most of that savings in the Bay Area. In addition, he said, it used to cost Sun $1,000 to move somebody into an office.

  (Johnson, Steve (2003, November 17). *Valley firms ditch desks to cut costs. Mercury News*)

Sometimes companies are surprised about how well a distributed-employee model can work. During the 1996 Olympics in Atlanta, people were requested by the State of Georgia to stay off the Atlanta roads as much as possible during the two weeks of the Olympic Games. HP's major site in Atlanta, which was very much in its infancy at that point as far as a telecommuting model, complied and asked knowledge workers, call center personnel, and the like to work from home during the Olympics. Amazingly, even in an era of 33k modem-access, support calls were still taken, financial analysis still got done, servers stayed up, and customers still had calls from sales reps. After that positive experience, some people started to ask, "Why do we need that big building?"

Chapter VII

Human Resources Advantages

Getting the Right Person for the Job Regardless of Location

Prior to the rise of virtual teams, hiring the right person for a job usually meant one of two things—limiting the pool of candidates or spending big bucks.

Unless a company is willing to spend to fly people in for interviews or to pay for company moves, it must hire within commuting distance of its facility. This may work out okay if you are in an employer's market and located in an area that has many people with the skill set for which you are looking (e.g., engineers in Silicon Valley). But what if it's an employee's market, and you're located in an area that lacks employees with your required skill set? Most likely, you'll end up compromising on what you really wanted in an employee. There is a big difference between the best available software engineer in Acworth, Georgia, and the best available software engineer in the Untied States.

The other non-virtual team option is to incur the expense of flying in candidates for interviews and (most likely) paying for their relocation, if they are from out of town (assuming you hire them, of course). We covered the joys of company-paid moves in the previous section.

If you're willing to hire a remote employee—either as a telecommuter or a worker in a remote office—suddenly, your pool of qualified candidates soars, and there is no $42,000 in relocation expenses.

Safety and Security

Two key factors for promoting the use of geographically dispersed virtual teams in the last several years are the threat of terrorism (9/11) and the threat of disease (SARS). Both have had a profound effect on how corporations view the co-located vs. geographically dispersed argument, as well as casting doubt on the necessity of virtual teams meeting face-to-face from time to time.

Prior to 9/11, it seemed a reasonable idea to group large numbers of people together in a single building. Many executives viewed this as superior to the idea of dispersing people among many locations, with efficiency, cost savings, and team cohesiveness among the arguments. September 11 radically changed that perception for many people. Just as consolidating your data center into a single location provides a single point of failure in case of disaster, so also does consolidating all of your people into a single location provide a single point of failure. After the two towers fell, suddenly people began to question the idea that co-located teams are always better. And as companies began to recover from the horrible destruction, it was virtual teaming and telecommuting that helped them get back on their feet (a similar situation existed during the 2003 wildfires in southern California).

In one case, a major technology industry analyst firm was able to publish several in-depth papers on the potential impact of the attacks within 72-hours of the fall of the towers. The papers were entirely written through a geographically dispersed team. Somewhat ironically, the same industry analyst has stressed in the past the importance of periodic face-to-face meetings for virtual teams.

The grounding of the U.S. commercial air system for several days after the attacks were also an eye opener for many people. It actually created a situation where it was impossible for many people to meet face-to-face, and it was a requirement that they meet virtually—an interesting juxtaposition. In the first year or so after the attacks, many people were hesitant to get on an airplane for fear of a terrorist attack. When confronted with the option of advancing their career by traveling or keeping themselves safe from terrorist attacks, most people choose the latter—and learned to live with virtual meetings.

Just as people were getting comfortable getting on a plane again, the SARS virus hit in early 2003 (the virus infected 8,098 people, and 774 died). One of the greatest risk factors for contracting the SARS virus was international travel by plane. Many companies temporarily banned travel to known infected locations (including China, Hong Kong, Tawain, and Toronto). People who were forced to fly to those locations (for customer visits, conferences, etc.), typically had a very queasy feeling about it. When one of our managers announced that he had to fly to Toronto for a customer visit during the height of the SARS outbreak, everyone on the phone became silent-no one thought that it was worth risking death for the visit. (Luckily, the visit was accomplished with no ill effects to our manager!) Just as with 9/11, the SARS epidemic changed the dynamic—where it was impossible for many people to meet face-to-face, and it was a requirement that they meet virtually.

Whether these events will have a long-term impact on corporate behavior remains to be seen. If the terrorism threat is wiped out, and there are no diseases *du jour*, perhaps people will return to their old patterns of traveling for meetings. But what 9/11 and the SARS epidemic accomplished was to prove that it *was* possible to get work done efficiently and effectively in virtual teams, without face-to-face meetings.

Providing Work/Life Balance

We would also postulate that working on a virtual team can help corporations provide a work/life balance opportunity for its employees. When one of the authors of this book returned from a maternity leave approximately five and a half years ago, she made a proposal to her virtual team manager (who was located at another site) for part-time work (using Hewlett-Packard's already established personnel processes for such a proposal). The author's proposal was approved, and she returned to work part-time and continues to work part-time as this book is being written.

In the author's experience, working on a virtual team has been a key enabler of the long-term success of this part-time work arrangement. One of the challenges (and, in some cases, complete roadblocks) that many people face when trying to make part-time work schedules successful in an office, team, or organization of co-located full-time workers is the fact that everyone is constantly aware of when the part-time employee is not in the office. Just a glance at the individual's desk serves as a reminder that he or she is "not working" that day, that hour, and so forth. Thus, the part-time employee is more likely to be labeled as "not available when needed" by their colleagues. There may also be underlying theories in circulation that the individual may not be "putting in his or her time," since they are not visible in the office as often as their full-time colleagues. This sentiment tends to bubble up to the management staff; the employee receives the feedback, and eventually gives up on the part-time arrangement—either by increasing his or her hours (even if he or she doesn't really want to) or by leaving the company.

However, in a virtual team environment, the potential negative dynamics associated with part-time work can be greatly reduced. Here are some of the reasons why, based on the author's experience.

- **Virtual team members are more likely to be evaluated based on their results.** As we discussed in Chapter 3, as corporations are moving more towards results-oriented human resources processes, employees are typically evaluated on the results they have achieved rather than the individual hours that have been worked. If employees are meeting their objectives, then they are likely to be successful. This type of results-oriented evaluation process is particularly needed within virtual teams, where many team members are located at sites away from their direct management staff and other team members.
- **Team members on geographically distributed teams are less likely to judge co-workers by whether they are sitting at their desks.** Again, when you have team members spread across the world, everyone is less likely to jump to conclusions regarding how much someone is working or not. If team members are meeting their commitments to you, then that is all that matters. Whether they are at their desk at 4:00 on Monday when you happen to walk by is irrelevant.
- **Virtual team members already must rely heavily on technology on a day-to-day basis.** Virtual teams must rely on technology (e-mail, voicemail, etc.) to be successful, since many team members are not co-located anyway. Much day-to-day interaction is already conducted electronically rather than by walking up to someone's desk to chat. While working part-time, one can still keep up on urgent e-mails and voicemails (even on days and during times that are not part of the official part-time work schedule), so the part-time schedule is less likely to impact co-workers.
- **Virtual teams tend to rely less on business travel.** Again, due to the geographically distributed nature of virtual teams, the teams tend to rely less on business travel for routine meetings, substituting electronic meeting methodologies instead. Part-time employees usually work part-time in

> support of other life objectives, such as spending more time with their family, pursuing a continuing education, and so forth. Therefore, they are less likely to be able to drop everything to travel to an out-of-town meeting. These types of last minute requests are typically less frequent within virtual teams, as any face-to-face meeting activities must be planned out well in advance to ensure maximum benefit for the travel dollars spent to gather the team together.

This is not to say that when you work on a virtual team it is easy to work part-time. Employees must still keep up on their commitments and achieve their results within their reduced work hours, make an effort to stay in contact through e-mail and voicemail, occasionally rearrange schedules as needed to attend to urgent business, and so forth. However, virtual teams and their methodologies of interacting can provide a more nurturing environment for the part-time employee.

Improving the Quality of Life

In addition to supporting part-time work, the distributed nature of virtual teams also has the potential to improve the quality of life for team members. Because team members can be located almost anywhere, you have the opportunity to live close to family or friends even if there isn't a local office nearby. You may have aging parents you need to stay near, close family ties you want to preserve, or a local doctor who you want to continue to see. Or you might have the opportunity to live in the mountains and be close to nature to continue your pursuit of hiking, fly fishing, and enjoying a climate you prefer. Alternatively, you may wish to live in the bustling city and be close to shopping and the arts. So you have a choice—be near mother, Mother Nature, or the Nature Company—it's up to you.

Another aspect of improving the quality of life is related to the variety of jobs available to you. Many jobs now take advantage of virtual teams and team members who are located away from the hiring office. This avoids the time and disruption required to pack up and move a household. There also may be more opportunities to explore different careers in different industries, because you are not limited to just the jobs that are close to you.

Blurring Work and Personal Time

When 8:00 A.M. to 5:00 P.M. was considered standard business hours, people woke up in the morning, had some personal time to themselves or with their families, and then went to work. They spent their workday in the office and then went back home to have more personal time before starting the next day. Working harder meant staying at work longer. Work/life balance meant setting aside enough time away from work to accomplish your personal goals.

The virtual team model and supporting tools have had an effect on how people manage their time and where they spend their time. In a virtual team, you spend most of your time either on the phone or exchanging e-mails or instant messages. So really, as long as you have Internet connectivity and a telephone, your physical location is not really as important as your ability to have connectivity. You could be at home, or you could be at a coffee shop with Internet connectivity. In either case, the need to be physically in the office is less of a requirement.

Work has become something that you do rather than a place to go.

Speed and flexibility increase greatly with virtual teams. If you don't need to spend time traveling from one meeting to another, especially if they are between campuses or cities, you can spend more productive time interacting with your team. Scheduling becomes easier, as well. You may have a face-to-face meeting with a customer in one location and immediately schedule a meeting with your virtual team on the phone during your drive to the next customer meeting. Or that face-to-face meeting might be with your child's teacher for a yearly parent/teacher conference. When you can be constantly connected via phone or Internet, interspersing business commitments with personal commitments becomes natural. Checking e-mail while at home in the morning, evening, and sometimes on the weekend is as common as taking care of short personal errands during the day. So today, rather than setting aside personal time and work time, they become interspersed.

Impact on Leisure Time

Although the blurring of work and personal time can have a positive effect on leisure time, primarily in the flexibility of scheduling activities there is a potential that the tools that help keep virtual team members connected are also the ones that begin to intrude into their personal lives.

People working on virtual teams develop habits of checking their voice messages frequently and constantly being available through e-mail, mobile phone, or instant messaging. It's especially tempting if you are working at home and the computer is turned on, or if you carry a mobile e-mail device to take a quick peek at your e-mail and send a reply to get a head start on your work day or work week. Because virtual team members may have developed the skill to intersperse work time and private time, it's tempting to tell co-workers that they can contact you on the weekend or on vacation. It requires discipline to manage your time so that job commitments and personal commitments can be met without one intruding upon the other.

Chapter VIII

Organizational Advantages

Tying Together a Decentralized Environment

In the mid-1990s, Hewlett-Packard was a very decentralized company, and IT was especially decentralized. At the time, HP had 150 data centers located around the world. Many of these data centers were owned and managed by HP businesses or factory sites, and only grudgingly did they interact with the broader HP. While there was a central corporate IT function located in Palo Alto, California, it only occasionally impacted the rest of the myriad HP IT organizations. Sometimes, that interaction was decreed (only TCP/IP would be used on internal networks, single e-mail systems, etc.), and sometimes corporate could only suggest (desktop application standards, PC hardware).

By the mid-1990s, one of the downsides of a decentralized IT function was becoming apparent—the TCO (Total Cost of Ownership) was tilting towards the management of PCs as the most expensive part, and not the hardware/software costs. But how do you manage 150,000 PCs in 120 countries, served by 150 data centers without any centralized function to mandate standards and management solutions? To solve this problem, one of the first (and greatest) examples of a virtual team in HP IT history started to be formed in 1994—the PC Common Operating Environment team.

While the PC COE team was chartered with coming up with software, hardware, and management standards for the company, we had no big stick to force anyone to follow our solution. We could only suggest, cajole, and sell our solution, and hope that the world (HP IT and businesses) would buy it. In the past, a team would have been formed in Palo Alto, and it would have been physically located in Building 20 at Corporate. Typically, the team would churn out white papers and standards, choose software, and so forth, which would have been greeted with a large yawn by the business/factory site IT departments, who received their funding from the business or site, not corporate. According to PC COE creator and founder Brandt Faatz:

> In general, I think that the site organizations looked at the centralized organization as a body that made recommendations—and they would generally take those recommendations to heart, but wouldn't necessarily follow them to the letter. They might adapt them, or they might even disagree with them, and go in a completely different direction.

So, how do you get everyone in the company going in the same direction when they don't have to? A unique, federated organization model soon developed. First, the people who originally architected PC COE worked for a factory site in Fort Collins, Colorado, not at corporate headquarters. So, to keep the original momentum and knowledge going, one of the Fort Collins people (Brandt Faatz) was made the head of the new team. Following the model of hiring the best people for the job regardless of their location, Faatz soon hired more people to the PC COE Core Team from HP IT sites around the world—Palo Alto, Boise, Atlanta, Bracknell, Melbourne, and Germany. This core team was not only virtual geographically, but also organizationally; many of the members were half-funded by the corporate PC COE team and half-funded by their local IT organization/business. Faatz comments on the partial-funding model:

> The reason that we went with the half-funded people is one, we wanted to secure enough resources that we had enough control or predictability to actually get the work done. The funded people tended to do specific work for us—they might be working on software, they might be writing up documentation—something to do with the next release of PC COE

> that we were working on. So we wanted to have a commitment from their manager that 25% of their time or 50% of their time was a formal commitment. The other thought was, and this is an extension of the extended team concept, you get those people funded, and they now become a focal point within that organization for additional resources to rally around. So we ended up getting a really good deal by funding these people, say half time.

An interesting thing happened as a result—suddenly this corporate IT team was made up of representatives from key IT organizations around the world, and all felt they had a stake in its results—and a voice at the table. This virtual team thing was paying off in more ways than just getting the best people for the job regardless of location!

In time, the model was so successful that a second tier was added—the PC COE Extended Team. This was comprised of representatives from IT organizations that didn't have people on the core team. While members of the extended team weren't funded by the corporate PC COE core team, they had equal voting rights and a place at the table for weekly and annual PC COE planning sessions. In time, the core team had about 20 people, and the extended team was closer to 50. Faatz comments on the importance of the extended team:

> They had a seat at the table, and a chance to be involved in the decision making, to understand the decision making. The extended team tended to be representatives from a lot of the major sites and organizations around the company. If we had to prepare the management or employees on a site for a particular change that was coming, we would call on them to help us with the management of change. We also engaged them

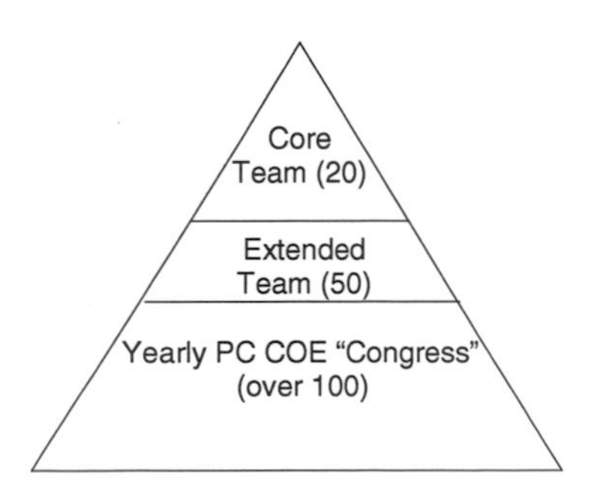

> heavily in the testing process. So they were already bought into the changes because they were involved in the decisions. They were very aware of the rationale that had gone into making decisions, and they had a vested interest in the success of the deployment.

Brandt also comments on another advantage to the PC COE virtual team model:

> The other thing that we could do with this whole structure of extended team, partially funded core team, and core team, it gave us an opportunity to manage the team composition a little bit differently than you would in a traditional organization. If we had someone in the outer rings of our model that was really engaged, really doing great work, we could get that person increasingly engaged, take advantage of them as a resource, at some point start funding part of their salary, and even at some point hire them into the organization. It was sort of a "try before you buy" model.

Finally, in days when travel budgets were a bit less strict, we'd have a once-a-year PC COE congress, where everyone on the core team, extended team, and business representatives would gather face-to-face to map out the high-level PC COE strategy for the coming year. The congress would often have more than 100 people in attendance and represented 30 or 40 IT organizations. In time, as technologies improved (and travel budgets tightened), the yearly congress became completely virtual and moved to the Web instead of face-to-face.

The PC COE project was one of the most successful IT projects in HP IT history. Within three years of its inception as a global program, PC COE was managing 150,000 PCs within HP, with an annual cost savings of $200,000,000 a year. And most interestingly from the point of view of this discussion, it was all done with a geographically and organizationally dispersed team.

The PC COE model was designed for a decentralized organizational model. Does it give us any lessons for a more centralized organization? Again, Faatz comments:

Our organization model and our decision-making model has changed pretty dramatically since 1995, and yet it still comes down to getting the right information, making sure you have a decision maker identified, communicating effectively, doing the management of change—all of that stuff is just as relevant in a more hierarchical, more centralized organization, but for slightly different reasons. Instead of needing everyone to agree because everyone has autonomy, you just need everyone to understand, and to make sure you're making the right decisions.

The idea is to be able to make decisions as quickly as you can while having enough information and enough exposure to the decision making process that you can easily make them stick. And it is very easy to fall on one or the other extreme, where you either have so many people involved and over-empowered that it's impossible to make effective decisions, or make them fast, or you have so few people involved and such poor communication that either you're making the wrong decisions because you don't have the right information, or you have a hard time communicating the decisions because nobody understands them once you do communicate them. So striking that balance is something that is a big challenge for us today.

Mergers and Acquisitions

Mergers and acquisitions in large corporations can be very painful propositions. In 2001/2002, HP was able to successfully use the virtual team concept to ease the complicated transitions required in the Hewlett-Packard/Compaq merger, which closed on May 6, 2002. The virtual teams, known as the "clean room," included representatives from both companies in the months leading up to the merger, and focused on specific areas, which included future product plans, organizational structure, and infrastructure. Sonja Poling, a member of the HP IT Infrastructure Planning clean room describes the purpose of the teams:

The purpose was to work together on plans for merging the company in a very controlled and legally acceptable fashion before the merger was approved by the SEC, the stockholders, and so forth. The clean

> room consisted of a small group of specifically identified people who could start planning in a confidential way for merging the company, and they were very limited in terms of what they could talk about.

The clean room was a classic virtual team both organizationally and geographically. People from multiple organizations and locations (spanning many nations and continents) in the two companies were brought together to prepare the way for the merger. The clean room was always meant to be a temporary assignment—once the merger was completed, most of the clean room activities would cease, and the virtual team members would disperse.

We interviewed a manager that served in the Personal Communication Infrastructure Strategy clean room. This group focused on long-term strategy in the areas of mobility, collaboration, cell phones, and so forth. The 10-member team had two members in Asia, one in Europe, and seven scattered across the United States. While many of the clean rooms had three to six months to complete their work, this one had a special challenge—completing their assignment in four weeks. As a result of this short timeline, the team never met face-to-face and did everything electronically.

> We never met face-to-face. It was an extremely short amount of time that we had—basically, four weeks. It was very, very intense. So we could not spend a lot of time traveling. We had to do very quick work. We couldn't afford the overhead of meeting face-to-face, so we did everything electronically. We did a lot of electronic meetings through NetMeeting, a lot of audio conference, and a lot of e-mail.

While overall the team worked well together, some cultural differences emerged between the working styles of the two companies.

> The company culture standpoint was primarily around this e-mail/voicemail-centric focus. HP, when it tried to move fast, used voicemail for offline communications. And Compaq had this very strong e-mail—and as well, terse e-mail—communication, almost to the sense of "offline Instant Messaging." You'd get these one-line or two-line e-mails, and that was the way they communicated. So we had to adapt to each other's style.

Another cultural difference seemed to be in the philosophical question of whether virtual teams were part of normal business, or whether they were something only used for small, one-off projects.

> People [team-members] were comfortable getting together in small teams and working on short projects remotely, but there was sort of a strong culture from a pre-merger Compaq standpoint that you don't do this as a normal part of your business. So, once in a while, if you have these sort of one-off projects, and you all had to be virtual, great. But this is not the way to run a day-to-day business. That's a cultural aspect that came out later, but it wasn't necessarily a hindrance to this particular team.

Key success factors of the team underscore the classic advantages of virtual teams—bringing together the best people from multiple organizations, on an as-needed basis, to accomplish a targeted task.

> The key success factors were, number one, having a very focused objective. Number two, was being able to draw upon the expertise of both pre-merger companies. We were able to get, basically, the thought leaders in this area independent of their physical location or organizational affiliation, and have them come into this team and contribute the strongest content possible.

How successful were the HP clean rooms overall? On "Day One" after the merger, the following was true:

- Detailed product roadmaps were available to customers.
- High-level organization charts were available within the company.
- Telephone systems, e-mail systems, networks, and the enterprise directory were integrated to the point where employees in the two pre-merger companies could easily communicate.

The HP/Compaq merger was also an interesting study in what happens when two companies merge—one with an extensive virtual team background (HP) and one that had a less extensive virtual team background (Compaq, which was very Houston-centric at the time of the merger). A team that we participated in was a microcosm of the bigger issue. We had 10 people—five located in Houston, literally within shouting distance of each other, and the other five (all pre-merger HP employees) located in Atlanta, Boise, Palo Alto, Germany, and England. The Houston-based team was used to face-to-face interactions. If the manager needed to assign something to someone on the fly, all he had to do was lean over his cubicle.

With the five remote employees, though, there was a long-time culture of virtual teaming, including extensive use of instant messaging, both for communication and presence information, as well as a highly developed set of virtual meeting protocols. So, the irresistible force meets the immovable object.

Initial team meetings were awkward. The Houston group met in a conference room and used a speaker phone/phone bridge for the remote employees. All of the problems that are inherent in that situation became quickly apparent—people on the phone rarely knew who was speaking, there was much side discussion among the conference room participants, and the quality of the speakerphone was abysmal. In time, our manager instituted a new policy—either we all meet face-to-face, or we all meet virtually. And since we were in the dot-com bust era of limited travel, that meant that most of our meetings moved to audio bridge/NetMeeting. In time, communication improved, and the two cultures melded together as one team. Instant messaging became the real-time communication path of choice and substituted for the water cooler conversations that the Houston group was used to with fellow teammates.

That team formed in July 2002. We wouldn't meet our manager face-to-face until Fall 2004.

Virtual teams turn out to be an excellent way of bringing disparate cultures together after mergers and acquisitions, as well as ensuring that no employees are at a disadvantage based on their location.

Getting Sales Representatives Out of the Office and in Front of Customers

In the early 1990s, Hewlett-Packard's field organization began a sales force automation program. Long before the technology had settled itself, sales reps were issued laptops, 28k modems, dial-in security devices, and sales force automation software. The objective—get sales reps out of the office and in front of customers. The theory—every minute a sales rep spends in the office is a minute they aren't in front of the customer selling. In time, for those that didn't get the message, HP followed up by removing individual office cubicles for sales reps and moving to a "hoteling" model, where reps shared office space on a first-come, first served basis. As the technology improved, sales reps were encouraged to work out of their homes, out of their cars, and at customer sites—anywhere but the office. Eventually, it wasn't unusual for HP sales offices to have less than 30% of the bays occupied on any given day. The result—sales reps in front of customers (better selling) and smaller HP offices (lower real estate costs).

Regional Representatives

Perhaps the most traditional use for virtual teams is to have regional representatives close to one's customer base. While many companies today, for example, have national sales teams for major customers, it is still often useful to have a local representative that can drop by the customer without travel or planning. Golf games, dinner, theater, and baseball tickets are still part of the personal selling process.

We know of a small company, Eddy Current Specialists, Inc. of Villa Rica, Georgia, that performs non-destructive testing on large air conditioners. ECS has followed this regional representative model for its technicians, with one or two technicians located in or near each state in which the company does business. Service orders come into the main office near Atlanta, and a regional technician is dispatched to do the testing.

We interviewed ECS President Ken Eisenhauer and Office Manager Debra Kasson. Ken describes two of the key advantages of having geographically dispersed technicians—price competitiveness and timeliness.

> It's cheaper—we can provide the service at a lower price, because people don't have to travel so far. We'll submit a bid to a company, and a competitor will submit a bid to the same company on a specific job, and because we're $50 cheaper because we're x-miles closer, we'll

> get it. And yes, they do work that tight! And timeliness is another advantage. If I was sitting here in Atlanta, and there was a job in Nashville, and we didn't have anybody in that area, I'd have to travel there. So, I'd have a day's travel time. Well, what if the customer wanted it done *now*? If I was there in Nashville, and I was available, I could knock it out. But if I was in Atlanta, I'd have a day's travel involved. So, it's customer response, too.

In order to make this geographically dispersed model work, ECS has digitalized and computerized as much of the analysts' reporting process (on the air conditioning units tested) as possible. Among the tools: Excel spreadsheets, scanned reports, ZIP files, and e-mail. While this digitalization of the process has been a boon to the efficient operation of the business, it has also upped the ante on the skill set required by the technicians. ECS's new technicians must be computer literate and willing to use digital means to create and submit reports. Eisenhauer comments:

> The digital age is a very good thing. Previously, everything was done by hand and mailed in by snail-mail. A machine may get done on the first of January, but by the time you get the report, maybe two weeks have passed. Then it has to be processed. And if there is a problem with it, you have to go back to them. So it might take up to three weeks before the customer gets the report in their hand. But now, if a machine is done on Monday, and everything goes well, we've got the report processed and out by Tuesday morning. We have competitors that it sometimes takes three months to do, and we just chuckle.

Kasson comments on how digital tools have helped make the geographic-dispersion model successful:

> With the common use of the cell phone and e-mail, working with a geographically dispersed group suddenly became much easier. Disseminating information prior to this was either done over the fax or by phone with the receiving person writing down the information. With the exception of one person, everyone has a computer (many also own laptops) and cell phones. Job information is e-mailed with follow-up confirmation done via phone. This is important particularly in the

> summer, which is our slow time of year, since an analyst may only work a couple of times a month. Without the "checks" in place a job could easily be overlooked.

ECS only has one non-digital person left in the work force, as Eisenhauer notes:

> We have a person who is not computer literate whatsoever, so he's doing his entire reports by hand—so to speak, the analog method—and they're being sent by snail mail. So, instead of everything being digital, we have to do photocopies of them, and because of that, we can't manipulate much of the data. It's locked in, because it's hard copy. But he's the last. Nobody else coming on will be allowed to use any kind of analog reporting.

Once a year, ECS has an annual meeting in the Atlanta area, where all members of the company are expected to attend. While this meeting is traditional within the company, Eisenhauer doesn't view it as a requirement:

> Actually, the way we're set up, there really isn't any need for a face-to-face at all, unless somebody has a specific issue that they want to bring up that they want to lay on everybody. But actually, we could get by without that.

Kasson agrees that the need for a face-to-face meeting has become less over time, but notes that there are occasional issues that require face-to-face interaction.

> It is only important to meet face-to-face under certain circumstances. The Board of Directors can meet via conference call. For the entire company, if there are stockholder issues forthcoming, changes being recommended to the company by-laws or something that will affect the company as a whole then a physical meeting should occur.

As a final thought on the ECS geographically dispersed model, Eisenhauer made a comment in his interview that could be the theme of this book.

> If I had to go without ever seeing anyone [ECS employees and contractors] again, it wouldn't hurt a thing—we could operate very effectively.

Innovation

Virtual teams can be very effective for driving innovation within an organization, team, or institution. In the same way that you can leverage a network of contacts to answer a question, you can leverage that network to create something new. With a co-located team, you're limited to resources and ideas that are local to you. Because they are local to you, they also share the same environment. They may have the same perspective on a problem or answer. They may not challenge what's considered a given. If the team has been together for a long time, they may even start thinking the same way. All of these behaviors could limit the amount of innovation and new ideas that the team generates.

In certain circumstances, it is difficult for the core team to be fully virtual. For example, in an R&D lab where certain pieces of lab equipment or research facilities are either very expensive or one of a kind in nature, the main team must be co-located with the equipment or facilities. However, even in these situations, they can supplement their research with a virtual team.

With a virtual team, potential contributors to the innovation process could be endless. It would not be out of the question to have an expert in a different country contribute ideas and input via e-mail or a phone conversation as part of a virtual team. But it would be highly unlikely that every team that wanted the expert's participation could afford to fly them out to be with their local team.

A virtual team also enables a team to bring in different perspectives that can help make the final result richer in content. Building a virtual team that includes people from different countries with different skill sets and different backgrounds brings more experiences into the discussion. Let's say you're in the United States trying to think of new ways to use a mobile phone. You might think about using it like a walkie-talkie or connecting it to your computer so you can transfer phone numbers between them. But if you had someone from Japan on your team, you might think about using the phone to pay for a candy bar in a vending machine.

Innovation in a virtual team might also happen faster, since you don't need to bring everyone to the same location before starting to work. You might also be able to leverage the fact that people are in different time zones to essentially have the innovation process working 24-hours a day instead of the typical eight- to 10-hour workday.

A great example of innovation through the leverage of virtual teams can be seen in the open source software community. This community is made up of software programmers throughout the world who volunteer their time and ideas for writing, modifying, and enhancing software that is freely distributed. The Linux operating system, Mozilla Web browser, and Apache Web server software that runs over half of the Web servers in the world, are examples of innovations that were created through the open source community. The open source community explains the benefits of its software development philosophy on its Web site at http://www.opensource.org.

> We in the open source community have learned that this rapid evolutionary process produces better software than the traditional closed model, in which only a very few programmers can see the source and everybody else must blindly use an opaque block of bits.

It is the teamwork, trust, and leverage of a broadly distributed team that helps make this community innovative and successful.

Other Virtual Team Drivers

In this section, we've discussed some of the key business drivers for virtual teaming—both in the geographical-dispersion and organizational sense. Here are some others that we would be remiss in not mentioning at least in passing:

- Better business partner collaboration
- Meeting federal rush hour commuting mandates for large companies
- Better regional representation on company projects

Why Some Companies Don't Get It

With all of the advantages to virtual teams, why are some companies more successful than others at their implementation? Why does it seem so natural in some organizations and so difficult in others? As it turns out, company culture can play a big part in whether virtual teams can be successful. Companies that have very hierarchical command and control organizations often will have greater difficulty implementing a virtual team model than those with a more matrixed, team-oriented environment. Hierarchical companies often create a very competitive atmosphere that discourages both team spirit and the free exchange of information. Companies with poor teamwork in general will definitely have problems with *virtual* teamwork.

Even companies that have enlightened views toward virtual teams still can have pockets of people that are hostile to them. Some typical reasons include the following:

- Managers like to be able to see people when they have assignments to hand out
- Managers like to be able to see if someone is at their desk and working
- Water cooler conversations are key to innovation and can't be replicated electronically

Indeed, managing a virtual team calls for increased focus on managing by objective (and judging the quality/timeliness of the results) and much less focus on when and where an employee works on the assigned tasks. This requires managers to plan assignments ahead of time and to be specific in what they are looking for. It is more difficult than assigning things on the fly to whomever happens to be sitting in their bay at the moment. Plus, there is a whole generation of managers who believe you seal the deal with a handshake and by looking someone in the eye. This is often impossible with geographically dispersed teams.

Section III

Virtual Team Models

In the Reader's Guide, we discussed our definition of virtual teams and presented an example. In this section, we'll explore some of the models for virtual teams. Specifically, we will review these virtual team models:

- **Mostly co-located with...**
  - One primary location, several remote team members
  - Several locations of concentration
- **Mostly virtual**

We'll track the progression of virtual teams from the more traditional model, where membership is often comprised of multiple co-located employees with only one or two individuals working in alternate locations (remote sites, telecommuting, etc.), to more recent models where team members are spread across different sites and different countries. We also will examine how the work methodologies and techniques employed by virtual teams have evolved over time. Virtual teams, for example, often have traveled for face-to-face meetings for functions such as project launches and checkpoint meetings. Today, however, as teams become more comfortable working virtually and as technology tools become more advanced, many virtual teams are now bypassing face-to-face meetings in an effort to speed business results, cut travel costs, and so forth. We also will introduce what we are labeling a "virtual team maturity curve" designed to help individuals or organizations determine how receptive they are to virtual work.

Chapter IX

Mostly Co-Located

We'll start with the mostly co-located model for virtual teams. In this model, the virtual team is comprised of one or two locations of team concentration (often with members working in the same building, site, geographic location, etc.), while there are several members working remotely from other places. This model is often how many organizations get their first start with virtual teaming, as some external event (a merger or acquisition, reorganization, desire to reward a favored employee, desire to get the right person for the job, etc.) spawns the need for more creative approaches to teamwork.

One Primary Location, Several Remote Team Members

In this mostly co-located model, there is one center of concentration of employees that is supplemented by several remote team members working in alternate locations (remote sites, telecommuting environments, etc.). This model often evolves to either reward a favored employee by letting them work at home or at an alternate location (probably the wrong reasons), or when a need arises for a specific skill set that cannot be filled locally. Often, with this model, the one or two remote employees are made to feel like the odd man or woman out, and team activity centers around the central "hive" where the bulk of the team members work. The remote team members "fly in" from time to time to the central hive for face-to-face meetings.

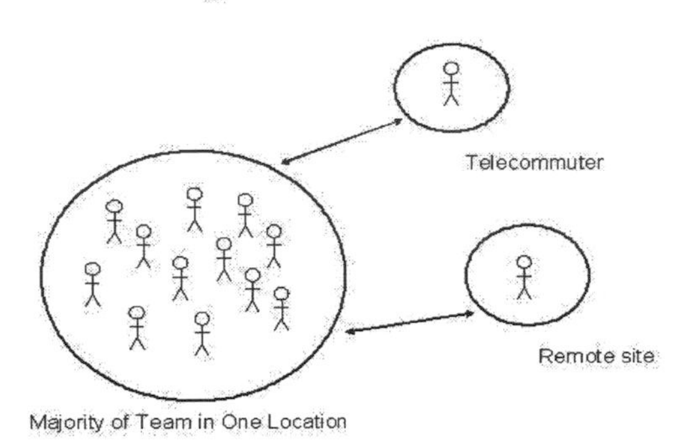

The model is particularly challenging for the remote members, as team interactions and work methodologies often are focused more on the face-to-face paradigms of the co-located members of the team, many of which we discussed in Chapter 4. For example, the process for routine team meetings often involves the centrally located staff gathering in a conference room while linking remote team members in via audioconferencing.

One early example of the use of the model within Hewlett-Packard occurred during the early days (early 1990s) of the aforementioned Sales Force Automation Projects. The teams that handled the project management and data center implementation role for the IT portion for these initiatives were located primarily in a central location in Atlanta, Georgia. However, hands-on implementation and project management coordination were provided by remote Technology Enablers (technology consultants) that were located in the field offices where sales representatives resided at the time. The Technology Enablers were often asked to dial into project-planning meetings via speaker phones (always poor quality, of course), while everyone else was gathered in a room together. Over time, however, the meeting methodology gradually evolved to where the entire group met exclusively via audioconferencing tools. At the beginning, it was a hard transition for the co-located bunch—meeting requests that were sent via e-mail without a specified conference room often received replies indicating, "You forgot to include the conference room in Atlanta for the meeting!" The reply back went something like, "No, we didn't—

we are trying a new way of doing things, and we aren't going to gather in a conference room!" With experience, the group became comfortable with the entirely electronic methodology for the meetings.

Several Locations of Concentration

In this co-located model, there are several locations of team concentration that are geographically distributed—perhaps half of the team works from a site in Atlanta, Georgia, and the other half from a site in Boise, Idaho. Sometimes this model emerges after a major reorganization or a company acquisition creates a need for groups at different locations to join together on the same team without the added expense of relocating either of the groups. Other times, it can evolve over time, beginning with more informal interactions between teams that eventually transition to more formal organizational ties.

Mostly Co-located:
Several Locations of Concentration

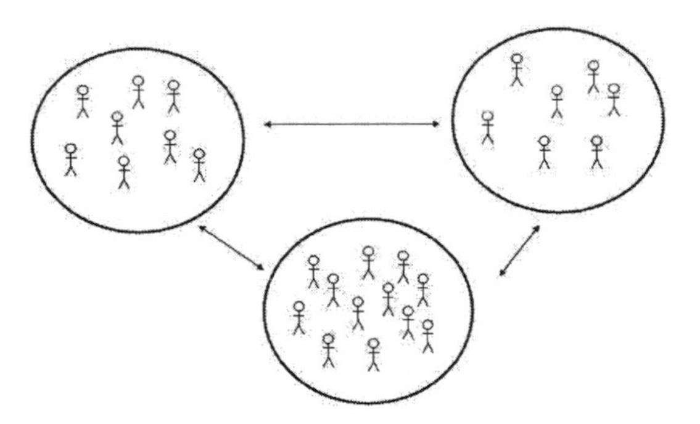

One early example of this model of virtual teaming within Hewlett-Packard is the previously mentioned Response Center organization, where geographically distributed employees were expected to work together to solve customers' technical problems. In the late 1980s, the customer support organizations within Hewlett-Packard were concentrated in several locations—the East and West Coasts in the United States, and sites in Europe and Asia-Pacific, as well. While each location initially had its own manager and team structure in place, customer support engineers at these locations were expected to work together, as needed, to solve customer problems on a 24/7 basis. The customer calls would "follow the sun," transferring (through a proprietary call tracking system) to the location that was best able to service the customers when they called.

Initially, in this early example, the individual teams providing customer support at the various customer call centers did not have regularly scheduled meetings nor did they interact in formal ways. Rather, they typically communicated electronically via e-mail or voicemail (the only tools available at the time). Often, messages also were exchanged within the call tracking system, which would follow the call as it traversed the time zones. Occasionally, the groups would participate in engineer exchanges, where staff from one location would travel to another to meet counterparts at the other locations and serve as a guest engineer on their teams for a week or two.

However, as the Response Center organization sought ways to improve efficiency and reduce costs, this more ad hoc model of interaction evolved over time, and customer support engineers began to work for remote managers.

Within Hewlett-Packard, this early experience in implementing virtual teams was certainly aided by investments that Hewlett-Packard had made in key IT infrastructures such as intranet, voicemail and e-mail.

It should be noted that some organizations actually may already be using a permutation of the mostly co-located model of virtual teams without actually naming them thus. Some examples might include the following examples:

- Suppose you have sales staff co-located within the three geographies of the United States (East Coast, Midwest, and West Coast). Both pre-sales and post-sales support teams are expected to collaborate to prepare customer proposals, support plans, and so forth, oftentimes working from alternate locations such as customer sites, home offices, and the like. Staff is expected to collaborate across geographies for major accounts that have a presence in more than one geographical area. Sales representatives and post-sales support share ideas, proposals, best practices, techniques, and so forth, via a team space located on a Web server.
- Another organization might have a product development staff co-located in Memphis, Tennessee, that creates the plans for new products. Manufacturing for the product is actually conducted in Grenoble, France. Online support for the product is centralized in India. What happens when a customer reports a problem to online support staff that is traced back

to a manufacturing problem requiring expertise from the original design staff to resolve? A team is quickly assembled with representatives from design, manufacturing, and support to help solve the customer problem. Given the geographic distribution of this organization, it is unlikely that all would travel to a face-to-face meeting, at least initially. Rather, they would use electronic tools, such as e-mail, audioconferencing, and the like to collaborate in order to develop a solution.

Each of these examples describes a form of virtual teaming that can occur within organizations.

Challenges to These Models

Of course, both of these mostly co-located models are challenged by the tendency of the co-located centers to fall back on non-virtual teamwork methodologies among those that are located together (e.g., "I prefer to have Bob on my project team because he sits right next to me, not Sam whose skills are better aligned with the need but sits in an alternate location"; "Let's all get together in a room to brainstorm ideas for the project," etc.).

In addition, it potentially may be harder to debunk the out-of-sight, out-of-mind dynamic if the management staff is concentrated only at the team's center of co-location.

Of course, one advantage to these two models is that if management has an individual project that they truly believe requires face-to-face interaction to be successful, they would be able to put together a face-to-face team from the locations of concentration, if needed, without incurring travel expenses.

Chapter X

Mostly Virtual

In the mostly virtual model, team members are spread around the globe. This model sometimes can evolve slowly. Perhaps a team starts with one or two members working remotely. Then, company reorganizations, mergers, or acquisitions may spark the need for additional geographically distributed team members. Alternatively, we've also seen cases where a team was formed to be mostly virtual and distributed from the beginning in order to meet a business need (the PC COC team we mentioned earlier serves as a key example).

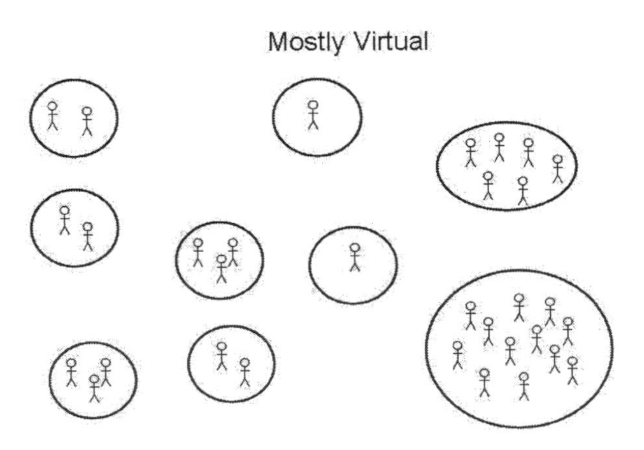

Advantages of a Mostly Virtual Team

Based on our experience, here are some of the advantages that the mostly virtual team model provides, compared to the mostly co-located models.

- There is a level playing field amongst all team members. Everyone is distributed and must rely on and use the tools available for team interactions.
- Provides fewer opportunities for what we often label center-of-the-universe syndrome. Center-of-the-universe syndrome can emerge when there is an area of large concentration of co-location for a particular virtual team. Team members in that location often develop the mindset that their location is the center of the universe and that any face-to-face meetings should occur there, virtual meetings should be scheduled to best accommodate that time zone, and so forth. For example, the PC COE team, a mostly virtual team, did not exhibit extreme center-of-the-universe syndrome. The periodic face-to-face meetings were held in various locations such as Fort Collins, Colorado; Atlanta, Georgia; Palo Alto, California; and so forth. Of course, the European and Asia-Pacific members of the team would probably describe the team as using the United States as the center of the universe.
- In this model, management staff is typically distributed, as well. This provides opportunities for everyone to interface face-to-face with management on occasion.

Challenges of a Completely Virtual Team

Of course, there are also challenges to working in the mostly virtual environment. Here are some of the key challenges that we've encountered. Several of them are discussed in more detail in our chapter on international issues, as they tend to be more prevalent as teams become distributed across time zones and cultures.

- Building community – Every team must be able to build a sense of community and common purpose among its members. Some techniques we've seen to address team building within virtual environments include:
  - Use of instant messaging
  - Frequent all-hands meetings (conducted electronically via Web-casting tools, etc.) to communicate items of general interest to the organization
  - Use of video, pictures, and the like, when available, to provide visual imagery
  - Occasional face-to-face gatherings to bring people together. (Note that although this is certainly a technique that can be used for team building, recent economic conditions have not favored these types of meetings, and they have declined in frequency.)
- Time zone challenges – In a completely virtual team, it is possible to have team members separated by 11 to 12 hours time difference. In some cases, time zones can be so disparate (Sidney, Australia versus Palo Alto, California) that very little overlap in reasonable business hours occurs. To meet together, someone may always be up late, up early, or working over lunch. The entire group needs to have empathy and understanding of this situation and rotate the inconvenience. Some techniques we've seen used to address this issue include:
  - Recording meetings for off-line listening. Basic technology is available today (software, hardware, etc) for recording meetings for off-line listening. Many audioconferencing services also provide recording as a value added service.
  - Offering multiple sessions of training, all hands meetings, and the like to accommodate disparate time zones
- Language and cultural differences can emerge – Clearly, language and cultural differences can arise among disparate team members from diverse backgrounds. In our experience, these issues are best handled on an as-needed basis. For example, language issues are often centered more on spoken interactions, and, in these cases, written communications can be more expedient.
- Attention span issues – Although a case could be made that all-day face-to-face meetings do not successfully hold an attendee's attention for eight hours, holding attention during lengthy virtual meetings can be challenging.

Until full immersion, 3D virtual meetings with holographic images are available, it remains to be seen how long one's attention span can be held through audioconferencing and data conferencing tools. We've found that three to four hours is probably the optimum length for a single session among virtual team members. (Virtual teams will typically run into time zone challenges when scheduling more time than that, anyway.)

Case Study: A Mostly Virtual Team Hosts a Summer Intern

> "I think that overall, the remote experience has been a very positive thing—the only aspect that has been lacking is a little bit of social contact." (Conor Gavin, 20-year old summer intern, HP)

We've seen mostly virtual team environments that conventional wisdom says are likely to fail, yet they've been outstanding successes. One example concerns Conor Gavin, a 20-year-old college student that did a summer (2004) internship with Hewlett-Packard in Palo Alto, California. The team he worked on—a network architecture team in HP's Managed Services Global Delivery organization—has employees scattered around the world, including the San Francisco Bay Area, Idaho, Munich, Corvallis, Vancouver, Grenoble, and Houston. Gavin spent the summer working as a remote member of that team. The only person he ever met on his team was his manager—and he only met her a handful of times. His mentor, Brian Jemes, was located in Idaho.

> On my first day I went into the office, and I met Mary my manager, and that was the first [face-to-face] contact I'd had with anyone on my team. And since that day, I've never met anyone else. And I've only seen Mary five or six times since then.

Jemes indicated that this was a new experience for the team—working with an intern that no one on the team had ever met (or ever would meet).

> It [mentoring a remote employee] did present a special challenge. We had talked about it—Mary and I—and said, 'Let's just view this as an experiment. Let's give it a try—theoretically, it seems like something we should be able to do, but we've never tried to before. And, we'll see how this experience goes.'

Conor describes how he communicated with his mentor (located in Boise) and how effective it was:

> I communicated with Brian more than anyone else on the team. In general, the way it worked was, I'd talk to Brian for a bit, he'd work through or explain how to do something, and then he'd leave me be for a day or two while I had a chunk of work to get through. And during that day-or-two time period, I'd use Instant Messenger if I had a quick question. I'd send him an e-mail with my latest progress.
>
> It was strange to communicate with people with nothing but e-mail, Instant Messenger, and telephone. I didn't have any faces to put with the voices. A little more personal contact would have been nice—the distance was a little impersonal. Now, having said that, working in that environment didn't bother me at all.

Conor describes the challenges—and advantages—of the remote model:

> I communicated with Brian regularly, but it still wasn't as much as if Brian was sitting in the cubicle next to me. I've talked to other interns—their boss was in the cube next to them. Their mentor is in the cube one down. They're constantly taking to their manager all day, every day. While with me, it's only when I need something, or I have a question, then I get in touch to talk. And if I have my work to do, and I have no questions, I might not talk to anybody for the entire day.
>
> It is a bit more relaxed without having a manager right there. Now, that's not a comment on Brian or Mary—they're very nice people. But you know how it is when your boss is constantly around you. It is more relaxed when I'm on my own. Some people work better with someone looking over they're shoulder—but I always found that I work better when I'm a bit more on my own. There is a fair amount of flexibility—there is a lot of freedom to work my own way.

Brian had some challenges to deal with in the relationship, too—while Conor had a background in physics and math, he had no networking experience. And HP was the first large corporation Conor had worked for—how could Brian integrate Conor into the company culture without being in the same geographic location as he?

> Probably the biggest challenge for me is that Conor didn't have networking background. He has a technical background with a physics and math background, which is helpful. That was my biggest challenge—how am I going to get him trained up? Because it just didn't seem like good use of our time to spend hours and hours on the phone. If I were there, I probably would have just pulled him into a conference room and white boarded stuff for a day or two, or a couple of half days or afternoons and done a lot of it myself. But, being remote, I looked at the "Learn at HP" site. I said, hey, that's a resource for a lot of training we get, and I wonder if there is some networking training out there, and I found actually there's quite a lot out there that no one on the team really knew was out there.
>
> I didn't really feel like we ended up struggling over any particular "passing on the company culture issues." I think he was personality-wise and temperament-wise a good fit with our team and the company as he came in. One thing that we did to help was, we had him go through our [Web-based] "Standards of Excellence" training. We felt like there was some amount of the company culture that we are trying to transmit to *everyone* through that training.
>
> We also encouraged Conor to get to know some of the other interns through the organized HP Intern events that were going on in the Bay Area. This enabled him to meet other HP people and get exposed to a broader range of the type of work HP is doing.

A key point about the overall experience is that the project that Conor was brought in to work on would have been significantly delayed if Conor and the team hadn't been willing to work in a remote environment, because it would have required extended travel and the accompanying expense:

> It was the kind of thing that basically would have been put on hold. It would have been very difficult to get approved travel if it was not

specifically related to a customer activity, and this was more for an internal IT initiative. I think we would have been limping along, and it probably would have been on hold for 6-9 months.

Chapter XI

How Teams Work Virtually

In this chapter, we will discuss some of the ways that virtual teams that we have participated in have worked together, and we will provide several case studies from our experiences. As background, it is important to remember that virtual teams can have several goals for operation.

- **Geographically distributed organization** – Here, an organization is composed of employees who are located in multiple locations; the team members share a common management reporting structure. From our example in the Reader's Guide, the Marketing, Development, and Online Support teams are geographically distributed organizations.
- **Geographically distributed project teams** – Here, a project team is comprised of individuals from different organizations (with different management reporting structures) who are brought together to deliver a specific set of results. The virtual project team often disbands once the team deliverable has been met; however, the authors have been members of geographically distributed project teams that were in place for multiple

years, producing ongoing deliverables in a specific area of interest. From our example in the Reader's Guide, the project team discussed would fall into this category.

Note that in a company that makes extensive use of the virtual team concept, individual employees can be in a geographically distributed organization while simultaneously participating in multiple geographically distributed project teams.

The Needs

While the needs of these two types of teams are very similar, given the nature of their makeup and goals, they do have some unique needs, as well. As this is not a book about generic management or project management skills, we won't try to produce an exhaustive listing, but here are some of the tasks that each type of virtual team must be able to accomplish. Although these needs are not unique to virtual teams (co-located teams would probably have the same or similar list), the needs are often more challenging to meet in virtual team models.

Geographically distributed organization	Geographically distributed project teams
o Conducting routine staff meetings o Building a sense of the team and the team's mission o Conducting larger organizational, all-hands meetings for updates and communications o Participating in human resources reviews as needed (personnel reviews, etc.) o Facilitating general communication among team members	o Building a common sense of purpose across divergent team members (who may also have divergent organizational priorities pulling them in alternate directions) o Securing project sponsorship o Conducting routine project meetings o Ensuring project communication

The Skills

In the beginning of the book, we discussed the essential competencies for virtual team members: communication and trust. However, there are other core competencies that virtual team members should have or develop. Many of these skills are enablers for collaborative work and can be applied to workgroups that are co-located as well as distributed.

Organizational skills, especially multi-tasking – Important in any situation, but essential when working in a virtual team, in order keep track of people that you've met, conversations that were had, and locating important pieces of information. With the blurring of personal and work life, managing activities and commitments becomes more complex. Even if you are able to separate your work life and personal life, your ability to manage information overload from multiple communication tools will be critical.

Time management – Especially during meetings, goals cannot be reached if discussions get side tracked and people start to disengage. In a face-to-face meeting, you can tell if people start reading their e-mail instead of paying attention to the discussion. In a virtual meeting, it won't be so obvious.

Attention to detail – Making sure you have gathered all your information and have had all your questions answered during a virtual interaction ensures that you don't have to frantically try to call the person back, try to set up another meeting, or interrupt them with an instant message. You don't have the leisure of finishing off the conversation while you walk back to your offices together.

Listening and testing for understanding – This is another critical skill, since almost all your interactions will be on the phone. All the physical emotional cues, except for tone of voice, are gone. You won't be able to see if the person is angry, embarrassed, confused, perplexed, or in disbelief. You'll need to listen closely to the tone of voice and what is said. When in doubt, probe for more input. Don't ask a group on the phone, "Do you have any questions?" because more than likely, there will be silence on the other end. Ask more direct questions like, "Bill, what issues do you see?" or "Pam, do you have any other suggestions?"

Summarizing and communicating clearly and succinctly – In a virtual environment, you don't have the luxury of being verbose or vague and letting the expression on your face or body language convey the rest of the message. You need to ensure that the words you write or speak are clear and unambiguous. If your message is not succinct, in a virtual meeting you will start to lose people's attention, and they will wander off and process e-mail or start surfing the Internet. Worse yet, a verbose and unclear e-mail could likely be ignored and filed in a "to be read" folder that is never accessed.

Empathy and encouragement – People who are not comfortable working in a virtual environment may tend to be hesitant to contribute or speak up during a conference call. People may not be as open to expressing their true view of things, because they feel they won't be able to judge the other team members' reactions. Silence during a conference call could be the result of those issues, or it might be that this is the third conference call the other person on the phone has had since 5:00 A.M. Encouraging people to contribute and trying to empathize with the other person's point of view can help in all of these situations.

Work in an ambiguous environment – As you can guess by reading previously about the core competencies, working in a virtual environment can be a confusing and frustrating experience. Your ability to work under ambiguous circumstances, and more importantly your ability to gain clarity under these conditions, will help you accomplish your goals and be more effective, even when you don't have all the data or all the answers.

The Tools

Virtual teams, of course, depend heavily on technology to assist them in their day-to-day success. Some of the tools that assist virtual teams include:

- Audioconference
- Instant messaging
- E-mail/voicemail
- Team spaces for document sharing, idea sharing, and the like (Groove, eRoom, etc.)
- Meeting management tools (Netmeeting, etc.)
- Web-casting tools (one to many communications, all-hands meetings, etc.)
- Desktop and room-based video

In addition, members of virtual teams sometimes will be asked to travel to periodic face-to-face meetings to supplement their electronic collaboration efforts. Some of the key reasons often cited for the need for these face-to-face meetings include:

- Project (or organizational) launches – The beginning of a project or the formation of a new organizational team is often cited as a time when team members should get together to meet face-to-face. That way, everyone can establish personal relationships (through more informal interactions over breaks, lunches, etc.), everyone can get on the same page regarding team or project delivers, reach agreement on goals, and so forth.
- Key project checkpoints where decisions will be made – The rationale for face-to-face interactions for these occasions is to ensure that all opinions are heard and assimilated and that everyone gives their buy-in for key decisions made (i.e., you must be able to look people in the eyes at these key checkpoints).

Of course, we are not postulating that face-to-face meetings of virtual teams are a waste of time. Rather, in many cases, they are a luxury that will not be affordable in the future. In today's business climate, organizations are under increased cost pressures that often limit business travel. World events (terrorist attacks, wildfires, SARs outbreaks, etc.) also can create situations where it is impossible to travel. In addition, the business world now moves at a faster pace, and the complex, logistical planning involved in bringing a group together face-to-face can sometimes exceed the length of time allocated for the team's deliverables. Imagine that you are a project manager in a customer service organization and are given this assignment by your management team:

> Our latest customer support surveys indicate that customer satisfaction is down over the last quarter. You have one week to prepare a list of at least 10 breakthrough ideas to present to our executive committee on how we can dramatically improve customer satisfaction across our customer support organization over the next three months. You must include participation from individuals within all five of our customer support centers—which are located in Atlanta, Georgia; Dallas, Texas; Munich, Germany; Paris, France, and Singapore.

Is your first step after receiving this assignment to immediately start planning a face to face meeting for all participants to brainstorm ideas? Probably not. Clearly, if a virtual project team is asked to prepare a report that will be available in one week, the team probably doesn't have time to coordinate a face-to-face meeting, particularly if the team has worldwide participation. Even if everyone on the team already had their plane tickets in hand, it would one to two business days of travel time just to get the team in the same location. However, the same results can be achieved through a virtual meeting using tools such as Netmeeting, audioconferencing, and a team space. An audioconference for all the players can be scheduled; the Netmeeting notepad can be used for all participants to brainstorm and prioritize the ideas formulated. The ideas can be captured and converted to a word processing document, which can be housed in the team repository during editing and refinement. The entire report is compiled and submitted to management within the one-week deadline.

Role of Video (Desktop and Room-Based)

In the experience of the authors, video has perhaps provided the least assistance to date in our virtual teams' interactions. Since one of the dynamics often cited that is missing from virtual team interactions is the visual queues that one receives during face-to-face meetings, conventional wisdom would indicate that video technology could clearly play an important role for virtual teams. In the authors' experience, however, the use of video has not been as successful as teams perhaps would like and has been somewhat limited in use. Of course, the limited use of video today is largely based on the quality of today's video technologies and technological complexity involved. If low-cost, broadcast quality video were available to the desktop today, most virtual teams would probably want to use it.

Note that many of today's business professionals may remember their first experiences with room-based video technologies in the 1980s. In those days, room-based video was often used to link together smaller groups from several sites for meetings. You may have had one group (10-15 people) in one location, with another group of the same size in a second or third locale. Technicians started an hour or more before the event to ensure that technical steps were taken to link the two (or three) videoconferencing rooms together, and they were lucky if a meeting ever started on time. Of course, the video quality was awful—characterized by slow motion movement, shadowlike figures lagging real time by at least three to four seconds, and so forth. The so-called visual queues that everyone was craving were lost somewhere along the way, anyway. In our opinion, these early experiences certainly have contributed to giving video a bad name, so to speak. When you mention video to many people today, they still oftentimes will remember these early experiences. We expect that with the evolution of broadband and improved quality, video will play more of a role in future virtual team interactions.

We have seen video used successfully in some of these areas:

- Interviews
- Customer presentations
- Web-casting one-to-many meetings (all-hands meetings, organizational updates, etc.)

We are also seeing younger generations embracing the video phone technologies, exchanging pictures and the like, so video may play a more prominent role in the future.

Virtual Teams Don't Play Putt-Putt Golf (or Informal Team Building on a Virtual Team)

When teams are completely co-located, there is often a quarterly or semi-annual effort to pursue a team building exercise where the group would gather for some type of non-work related activity, such as putt-putt golf, bowling, laser tag, and the like. The thought behind such events was that a team that plays together, works together better, and that these type of casual, informal interactions would enable team building. Even as virtual teams emerged, when the teams would meet face-to-face, time was often set aside in the meeting schedules for these types of activities. While we are not opposed to teams having fun together (and the correlating impact that these casual, informal interactions can have on business relationships), it should be noted that some team members actually may have resented these types of activities on the grounds that they often encroached on personal time, they were embarrassing (not everyone is a good bowler), and so forth.

Of course, building some sense of interpersonal relationships can assist virtual teams, so here are some of the ways that we've seen that effort made.

- Personal Web pages – Personal Web pages on the company intranet can be utilized to share highlights of personal activities (club memberships, civic projects, family activities, hobbies, etc.) to help others get to know team members better, provided that the activities cited do not conflict with company standards of conduct. (For example, this might not be the place for someone to reveal that they are the webmaster for an offshore gambling Web site.)

  One creative example we've seen is where an entire team posts pictures and brief biographies together on a team Web site. The biographies (which, in this case, were written by other team members who interviewed the individual—another opportunity for increased team interaction) can include some personal tidbits that team members might be comfortable sharing and would be the type of information that might be exchanged more informally between face-to-face team members at the water cooler, over lunches, and the like.
- Personal sections on status reports – We've seen team members occasionally share a brief personal tidbit by including it in status reports in a section marked "personal." Team members, for example, can share significant personal milestones (this week was my 15th wedding anniversary), personal accomplishments (I successfully completed the Boston Marathon this week), and so forth. In the co-located team model, this is roughly analogous to the type of informal interaction that might take place at the water cooler or coffee station, or that might be prompted by observing photos displayed at a colleague's desk or discussed while playing putt-putt golf.
- Instant messaging – Instant messaging tools also can be utilized to help enable the informal chatting that might occur at the water cooler or in the hallway ("How was your weekend?" "Good morning." "I'm signing off, talk to you tomorrow," etc.).

Some tools for informal team building on a virtual team:
- Personal Web pages
- "Personal" sections on status reports
- Instant messaging

Less "Clique-ish" Behavior

With virtual teams, it can be much harder to have cliques develop. In a co-located team environment, there are inevitably smaller groups that will have common interests outside of the workplace (hobbies, clubs, sports, etc.) and get together during off-work time to pursue these interests. Of course, on Monday morning, when the majority of the group is discussing how much fun they had playing football in the park over the weekend, it's hard not to feel left out. In the case of virtual teams, this type of dynamic is less likely to occur, since everyone is distributed around the globe.

Case Studies Illustrating Creative Ways to Address Working Together Within Virtual Teams

Here we will highlight several case studies of ways we have seen virtual teams creatively address situations that normally would have been handled face-to-face for co-located teams, including:

- An organizational "farewell" party
- A team building/celebration event
- Determining what projects an organization will pursue
- Substituting for a multi-day, face-to-face meeting
- Getting quick, routine decisions made
- Facilitating informal interaction among an entire virtual team—the team "water cooler"

Organizational "Celebration and Farewell"

Here is a case study from an organization called ISE that was disbanding due to company reorganization efforts. The organization of approximately 90 individuals was disbanding, and the team members were moving to other teams. The management team wanted to have what they labeled a "closing ceremony" for the group to provide closure for the organization and to celebrate the successes of the team. To give an idea of the geographic distribution of the team, here is an approximate summary of the members' locations.

Location	Number of Employees
Atlanta, Georgia	10
Roseville, California	6
Boise, Idaho	5
Fort Collins, Colorado	10
Palo Alto, California	37
Corvallis	5
Rockville, Maryland	2
Europe	6
Singapore	1
Other areas	7

The focus of the disbanding organization was what some analysts would describe as an advanced technology group, chartered with conducting information technology research and strategy. As such, the group embarked on a quest to provide a virtual celebration that would illustrate what could be done with the technology available.

Loyal Mealer, a manager within the organization as well as one of the virtual celebration planners, describes these goals for the virtual celebration:

> We were trying to achieve two things. First, provide closure [for the members of the organization]. And second, to test whether an event like this could be done virtually.

Carol Wolf, one of the event planners, elaborates on some of the dynamics that went into the planning:

> At one extreme were people who envisioned the event as a number of simultaneous but independent parties. They wanted people to be able to see videos of the parties taking place elsewhere, but assumed that all the individual interactions would take place face-to-face at each site. (Think of a New Year's Eve party where people occasionally glance at a television to see people simultaneously celebrating in other places.) Most of these people were at sites with many ISE people and had a history of hosting on-site parties.
>
> At the other extreme were people who envisioned the event as one worldwide party in which everyone would participate. They assumed that through the activities, people would get glimpses of, and even some fleeting interactions with, their colleagues at other sites. Most of these people were at sites where there were just a few ISE people, or where they were the only person from ISE. They were elated to be included, because, in the past, they would receive invitations to "all-ISE" parties held at the big sites, but there would be no provision for them to attend remotely and no budget to travel. This time, there was provision for remote attendance, and that in itself was very encouraging.

Loyal describes the planning process for the event.

> The planning process was done, of course, virtually. We did not have a great deal of time to plan, since things were closing down pretty quickly. Travel was not an option due to cost and time. We considered dozens of ideas for this.
>
> Meeting 100% virtually with everyone at their desks was rejected, since many people wanted more personal interactions (face-to-face) for those that could do it cheaply;
>
> Video was rejected due to the quality issues and the technical complexities of setting up the event. We did have a small group of people put together a retrospective in pictures and music that was extremely well received. We ran it over NetMeeting.
>
> Coordinated food, again, rejected from a cost and logistics standpoint. We opted instead for site-specific catering (minimal) and care packages for those who could not attend a local venue.

> We considered a large array of virtual games but settled on a simple Pictionary derivative optimized for NetMeeting. Logistics, training, existing tools, experience, planning time, etc., were the reasons for going simple.
>
> We also considered the length of the event everywhere from 90 minutes to all day. We settled on about 3-4 hours.

Again, Carol elaborates on the challenges during the planning process:

> We knew that the toughest events to make successful are hybrid meetings like this, with most participants face-to-face in one place, some participants in smaller face-to-face groups in several other places, and the rest participating as isolated individuals. We briefly considered trying to level the playing field, either by bringing everyone together at least in clusters, or by having everyone participate virtually from their desks. We quickly abandoned that idea because there was no money for travel to bring remote people together, and, realistically, we could not have persuaded the ISE decision-makers to give up the face-to-face aspects of the event.
>
> We knew, therefore, that the celebration experience would be very different for the various groups of participants. We encouraged the committee to do three things: first, to provide a basic theme and structure for the event which could be adjusted to fit the various participant groups; second, to provide a ceremonial section of the event (the "rite of passage" section) that was accessible to all participants; and, third, to structure the game competition section to be as versatile as possible, for accommodating people's conflicting party goals. The technology came into play in support of those three things.

Ultimately, the group held an approximately three- to four-hour virtual celebration across all participating sites. At locations where there were clusters of team members, the members gathered in a room together. Refreshments were provided at each location. Through data conferencing technology, the individual rooms were linked together. A slide show highlighting the successes of the organization was presented by the management team, and the group even played a form of picture-based charades, using the data conferencing tool's white board feature. After the virtual celebration, a Web site was utilized to showcase the slide set and pictures from the individual sites that participated. Carol describes the various technologies that were utilized during the celebration:

> We used e-mail to communicate with participants before the celebration. For the event itself, we relied on electronic imaging (including video, still photos, and drawings), print technology, voice conferencing, and shared whiteboards.
>
> Video cameras were simply set up to show the parties at each of the larger sites. Still photos and drawings conveyed the "beach and surfing" theme of the celebration. There was an amusing "rogues gallery" showing faces of everyone in ISE attached to cartoon bodies in a series of beach scenes. Even the agenda was graphical, showing different agenda items scattered on a sandy cartoon beach, with little barefoot prints leading from item to item. All the sites, even isolated individuals, had access to excellent HP color printers, so some sites printed the items as posters and hung them on the walls, while others distributed them as handouts.
>
> Voice conferencing linked everyone together. The large sites used high-quality speaker phones (in some cases with remote microphones for people making speeches), and most people dialing in individually had telephones with comfortable headsets and "mute" buttons. When the time came for the game competition, we used the "shared whiteboard" feature of HP Virtual Classroom, so as one person drew on the whiteboard, his or her teammates and the opposing team, regardless of where they were around the world, could watch the drawing take shape and call out their guesses.

Was this exactly the same as getting all 90 people in a room together for a celebration? Of course not, but it did enable the team to get a sense of closure before moving on to their new responsibilities. Loyal describes some of the feedback received:

> The event was extremely well attended and received.... The most common theme from the feedback received was that people were glad

> to have had the chance to bring closure to the organization. They were amazed that the virtual event worked as well as it did. Skeptics thought it could not be done. Surprisingly, one of the more well received aspects was the photo and music collage. People, for the first time, saw pictures of their colleagues with whom they had been working for many years. This is really indicative of just how virtual HP has become.

As video conferencing and other technologies continue to improve, the quality of such interactions only can improve. In addition, Carol also shares thoughts on the role of technology in these of events, noting:

> Technology isn't the limiting factor in remote events anymore. If you're clear about what you're trying to do, the technology is available to do it. In fact, usually it can be done with much simpler technology than people expect. But success requires clarity of purpose, and ingenuity to be able to accomplish familiar things in new ways. And one of the biggest traps, the deadliest pitfall, is to have a double standard for measuring success. Everyone knows that face-to-face meetings can be frustrating, face-to-face speeches can be boring, face-to-face ceremonies can seem to go on forever, and face-to-face parties don't always please everyone. But when they think about holding a virtual event, they don't think, "Can we do this virtually as well as we could do it face-to-face, good parts, boring bits, and all?" Instead, they either think that they shouldn't have a virtual event unless every minute will be effective, exciting, and enchanting for everyone, or—even worse—they expect that just because it is virtual, it will magically become effective, exciting, and enchanting for everyone, and then they blame the technology if it isn't.
>
> Holding events remotely can give people a fresh look at what they're doing and why they're doing it. We don't just want virtual events. We want better events, which we want to be virtual, when needed. The celebration succeeded because it was a good event, good enough to handle a lot of different expectations and still be worthwhile. Technology is one empowering piece of the whole, and although it isn't magic, it's a piece that has come of age.

Periodic Organizational Get-Togethers/Team Building

Earlier on, the same organization held an ice cream social for team members that were located at one of the central sites with a higher concentration of employees. Instead of saying, "Well, you don't work at the home office, so you lose out," the management team came up with a creative solution to help remote team members feel included in the site activities. On the same day as the ice cream social, all remote team members received a package (via interoffice mail) that included a T-shirt (that was handed out at the party) as well as gift certificates to purchase an ice-cream cone at a national ice-cream chain. Again, not the same as being there, but remote team members could still feel a part of the proceedings.

Determining What Projects an Organization Will Pursue

A geographically distributed team is often challenged by keeping participants involved in what projects are being pursued within the organization, who is managing those projects, and how one gets a project done on that list of key priorities. To address this situation, one organization that the authors worked for conducted quarterly sessions that were called "VC Cafés" (VC was an abbreviation for "venture capital"). The format was a two- to three-hour audioconference that was open to all members of the 90+-person organization who wished to submit a proposal. Prior to the audioconference, each of the organization's individual teams (through their own individual virtual staff meetings) culled through lists of potential projects that they would like to see funded by the management staff, selecting three or four of the top projects to present during the VC Café session. The 15- to 20-minute pitches for the projects (why they were important, what business need they met, what resources would be needed, etc.) were presented to the management staff during the VC Café. The management staff then selected the top projects for funding (through an off-line voting mechanism), indicating that those projects could move forward with implementation. All this was done with full visibility and participation of the entire geographically distributed organization.

Substituting for a Multi-Day, Face-to-Face Meeting

Many teams eventually will discover that they have a lengthy list of key issues that they need to discuss and resolve that may each take two to three hours (or more) to explore in detail. Perhaps it's time to do in-depth work on the team's charter or outline detailed goals for the coming year. This is a time when many might say, "Let's put together an agenda and get the team together for a multi-day face to face meeting to go down through the list." That's all well and good, until economic conditions prohibit business travel for internal meetings or world events dictate that travel to certain areas is restricted. Then, what do you do?

One team that two of the authors have worked on took this approach. A list of agenda items that the team needed to discuss was assembled and prioritized. The top priorities were gathered, and then a series of three- to four-hour audioconference sessions (facilitated by meeting management software) were scheduled across multiple days to address each of the items. Some of the sessions were completely devoted to one item, while others handled two or three, depending on the amount of discussion needed. In this way, each of the items got the full amount of attention needed, and the sessions could accommodate all geographically distributed members (both in the United States and Europe) with only marginal time zone inconveniences.

Getting Quick, Routine Decisions Made

Here's a recent experience that one of the authors had when a need arose to have a quick, routine decision made that required input and approval by a manager. I left a voicemail message for my manager highlighting the situation and offering several of the options we could pursue. Within about 20 minutes, I received an instant messaging message from my manager indicating that he had received the voicemail message and had some thoughts on the topic. We spent about five minutes exchanging ideas via instant messaging, and then the decision was made. I was able to provide an answer to the individual who had asked me the question within about 30 minutes—all facilitated electronically. (Of course, in some cases, one might desire a permanent audit trail of this type of activity.

It would have been very easy to save the instant messaging exchange and then forward a summary in e-mail to provide a permanent record of the decision making.)

Facilitating Informal Interaction Among an Entire Virtual Team: The Team "Water Cooler"

Virtual teams must find a way to simulate the casual conversations—often called the "water cooler"— that occur among members of face-to-face teams. We've already discussed how Instant Messaging has provided a valuable tool for this type of interaction among individual members of virtual teams. Most Instant Messaging products allow users to add additional people to their conversations, so three or four team members could be involved in the same chat session.

However, what if you want to create an environment where your entire virtual team could gather around the "water cooler" for an hour or so each week to share more informal information than would be normally discussed during routine staff meetings or project update meetings? We've seen a team develop an approach to do just that by holding weekly virtual meetings that they call "roundtable discussions." Lyle Harp, the group facilitator, shares the goals of the roundtables:

> The goal is to replicate the casual conversations that a co-located team would have naturally over the course of a week. This casual exchange would involve things as simple as overheard comments or the delivery of a box. For example, with a virtual team, the arrival of new servers must be announced, while a co-located team can see the stack of cardboard boxes. Many of the simplest clues concerning project and team status are lost when a team goes virtual. The other thing is brainstorming and group input. Where team members might discuss an idea of cubical walls over lunch, virtual teams need to schedule specific times to encourage this interaction.

Lyle explains how the sessions work:

> All Roundtable sessions are open to the whole team. We use a phone bridge and NetMeeting or HPVC [HP Virtual Classroom] for sharing documents, demos, and a whiteboard. A list of attendees is put up on the whiteboard as people join the meeting, and any notes that anyone feels are relevant. Every other meeting is spent giving verbal status reports on current projects. The format is open and informal. It's a good time for people to ask questions and find out what's going on with teammates' projects. On alternating weeks, specific topics are addressed where team members spend some time brainstorming and gathering ideas specific to a current project, proposal, presentation, etc.

Lyle shares these learnings from his experience:

> Virtual teams need to have "planned downtime" where job-relevant topics are discussed and kicked around as a team. Otherwise, the team becomes a group of isolated individuals.

Chapter XII

International Issues

The teams that we've worked on in the last 10 years have all been international in scope. And being a member of a virtual team for a United States corporation, but being physically located in another country, can have it's own set of problems, including time zone differences, infrastructure challenges, cultural issues, and language barriers. From an international perspective, we interviewed individuals in Australia, France, and Japan. We spoke with two members of the Hewlett-Packard PC COE team (discussed in Chapter 8), Geoff Markley, from Australia, and David Brehm, who spent part of his PC COE years living in California, and part living in France. We also interviewed Takehiko Kato from Japan who came to work in the US for four months in order to build stronger ties with his teammates and experiment with technology that was not yet available in Japan.

Australia

"In Asia-Pacific, the economics of flying everyone around every second week is prohibitively expensive. As a result of that, virtual teams have become a lot more commonplace."

Because of Australia's location in respect to the US and the rest of the Asia-Pacific area, Australia has good reason to invest in virtual teams and meetings. Geoff Markley, based in Melbourne, Australia, comments:

> The first is, if you're going to do any kind of global job, Australia is the worst place in the world to be, from that perspective. The closest place is Singapore which is 7 and a half hours by plane, and the U.S. is some large amount of hours after that. So in terms of when you physically had to be places around the other side of the world, there was lots of travel.

However, working on an international virtual team while residing in Australia does have its challenges. The first and foremost: time zone issues:

> If you're going to be working in virtual teams, especially in Australia, working in a regional or global role, you tend to be working non-standard business hours, just because of the range of time zones across even Asia Pacific. You start at New Zealand, which is 2 hours ahead of the East Coast of Australia, and you go through to India, which is 4/5 hours (depending on the time of year) behind Australia—even that is a very wide time zone range. And then mix Europe or the U.S. into that, and you end up working very strange hours. In the case of the U.S., when Daylight Savings changes twice yearly, there is a 2 hour time shift between Australia and the U.S., and so meetings go from starting at 11:00/midnight at one time of the year, to starting at 2:00am/3:00am for the rest of the year.

Because of the large number of Asian languages, language barriers are often an issue for virtual teams in Asia-Pacific, a problem that has improved in recent years as English has emerged as the default language of IT.

> The road to integration of virtual teams is certainly heading in the right direction as it becomes much more of the norm, and the fluency in English in a lot of the countries such as China and Korea where English fluency wasn't high a number of years ago is certainly helping that, and creating the ability for virtual teams, because the only common language in Asia is English.

Cultural barriers for virtual teams can be an issue also, especially in countries such as Japan which have a strong face-to-face business culture. The compromise seems to be having face-to-face kickoff meetings (and quarterly or checkpoint face-to-face meetings thereafter) and meeting virtually the rest of the time.

> If we look at a country like Japan, if you're doing business in a traditional Japanese style, absolutely you have to go meet them face-to-face—that's still very much prevalent. I think the IT industry is a bit of an exception to that, because of the requirement to have English as a fluent language—it is the international language for IT, and because there are a large amount of multi-national companies like HP. It's a little more nontraditional in some of these countries.
>
> I think, however, it's still important to meet the people, if you're going to have a long-term or a very detailed engagement with people, I think it's still critical to have fairly regular face-to-face meetings, and regular may depend on how important it is. For example, for the telephony company transition that we did last year (2003), I traveled once to Malaysia, once to Japan, and once to China. The rest of the time it was basically telephone conferences.

Dealing with the US can have its own set of virtual teaming issues. In the mid-1990s, during the PC COE team years, the problem was cultural awareness.

> In the early days of PC COE, my experience was that the people on the core team were quite good, but a lot of the ancillary people, whose role really was U.S.-focused, and were suddenly thrust into a global context, really were out of their depth. I'm sure in the early days there were lots of people [Americans] who had no idea where Australia actually was, and what the whole "working outside of the U.S." kind of concept was. So, by force or by stealth, I educated them, so that was also a challenge.

The situation has improved in recent years.

> I think a lot of the U.S. folks that I've worked with in more recent times have become a lot more understanding and knowledgeable of what happens outside of the U.S. in terms of working with virtual teams.

In the mid-1990s, there were often infrastructure barriers—poor telephone connections, lack of data conferencing tools, and the like. Geoff comments that this situation has improved dramatically in recent years, and notes two especially important advancements in virtual meetings—e-mail scheduling, which automatically converts the meeting times into the recipient's time zone, and local country dial-in numbers for phone conferences.

> I think the advent of the virtual technologies—things like Virtual Classroom, NetMeeting, these kind of virtual collaboration technologies certainly helped the ability of the virtual team meeting to be productive, with the combination of very good audio conference facilities and then the type of WebEx/Virtual Classroom-type facilities has made face-to-face meetings and training not very necessary.
>
> The telephony infrastructure for Asia-Pacific and the quality of the telephone lines has improved markedly. The distribution of the audio conferencing service that HP uses, where there are toll free numbers in every country, makes it very simple for people to dial in. The availability of calendaring in Outlook to schedule the meetings regularly with the people so that people know where to dial in and when to dial in has made a huge difference.

Geoff also notes that work/life balance issues can be a significant driving force for meeting virtually.

> In the early days of PC COE, when I was young and single and unattached, travel was sort of fun—"when's the next plane?" Now in 2004, I'm married, have two young children, so it's a little harder to go away for extended periods of time. You start looking for a lot more of those virtual options as opposed to having to jump on a plane every second week, which is very tough on the family, having to be away for so long.

Europe

David Brehm served on the PC COE team, living both in California and France. He later served as manager of a multi-European team of 200 IT infrastructure specialists. He has a unique perspective on what it's like to serve on a virtual team in both the U.S. and France.

Like Australia, there are time zone issues.

> It's interesting, because I'd been on both sides of the pond with the same team. I'd done the team from the States, where your morning started early—you'd usually have a conference call at 8:30a every week, and suddenly the conference call started at 5:30 in the evening. Instead of getting up early, I found that my day really started after the U.S. "got up" at 5:00p. So, it just sort of shifted things. You'd go around thc officc, and all of the people in global IT would be working late.
>
> Sometimes the U.S. people would schedule meetings and have no concept that some of the people on the phone were in other time zones, so we'd be on the phone at 11:00 or 12:00/midnight. It made it difficult to be there for "down time", if you had kids.

Also like Australia, there are language barrier issues.

> In the European context, you have all of the countries in Europe, so that added another layer of complexity that you had to be aware of. You had language differences, so it was difficult sometimes to understand what people were saying. One time I ended up acting almost like a translator between two people on a telephone conference, even though they both spoke English. One of them was from Atlanta, so he had sort of a Southern accent, and the second was from France, and they just could not understand each other. I had to translate—what *he* said was this, and what *he* said was that—it was just kind of funny.

There are also cultural barriers to deal with, especially regarding work styles in different countries.

> I had a German, a French, and an Italian manager who all worked together, but they came at the problem differently. The Germans I worked with were really into detail—they wanted all the facts. They were always asking me for spreadsheets. It's the German engineering mentality. And in France, I had a French boss, and he was really great at thinking about the "model" and the "big picture"—he was very intellectual. He was focused on how the pieces fit together, and very creative. And then I had an Italian boss who was very emotional, very passionate about the work. And to try to see the three of them work together—they were coming at it from all different angles—it was kind of challenging to watch them work together.

David soon found that he had to adapt to local customs to get things done, as this anecdote regarding getting plane tickets shows.

> I followed the process to get my plane ticket, because I traveled all the time over there [Europe], and I could never get whatever plane ticket I needed quickly enough. I was going through the process, and it just wasn't working. So, finally I went to the travel department face-to-face, and I said, look, I'm going to the U.S. I'd be happy to pick you up something—is there anything you want? I made friends with this one travel agent. Her boyfriend really liked maple syrup. So I brought her back some maple syrup. She was so happy—I think her boyfriend ate the whole thing the next day. And she was so happy about that that I never had any problems traveling for three years after that.
>
> That was the correct approach to use in France—you really need to establish relationships with people. They need to be able to trust you, and it's all based on a relationship. Whereas in Germany, it's all about the process. So I think that in countries that are process oriented, being remote is not a huge deal. Being in countries that are very relationship focused—Italy is that way, France is that way—face-to-face can be much more important.

As is the case in Australia, virtual teams and virtual meetings in HP Europe have become the norm, rather than the exception, in the last several years.

Japan

In 1989, Takehiko Kato was a network engineer working in Japan for HP. His senior management team decided to send Kato-san to the US for four months. Kato-san describes the objective of his trip:

> The purpose of my trip was to understand and experience the latest IT technology which was not well introduced or implemented but had demand or planned in Japan, and to create the good relationships by knowing each team and each team member as much as possible so that we can cooperatively work in the future. Another purpose was to improve my communication skill in English.

Kato-san echoed Geoff Markley's comments regarding virtual teams in Japan.

> In general from a Japanese view (but not from HP Japan's view), the virtual team is not so popular since in Japan the team is tightly aligned to a local organization. However, around the time I visited the US, extended business trips and foreign services employees (long-term transfers) from Japan to other countries were common.

Although there were some challenges during Kato-san's visit, the face-to-face interaction with his teammates in the US helped him feel more integrated into the team.

> Personally the extended trip was my first business trip, so I was nervous about many things not only job related but also my personal life including communication in English. The extended stay gave us much opportunity to understand each other to establish more strong friendship or relationship, also to understand culture or environment and it makes us more easier to know why they think a certain way, or how they come up with such an idea or opinion. Of course I was a visitor, but I was feeling as one of the team members. I was very motivated to feel that I was a part of the team and I think it became easier to implement new things supported by the US team after my assignment.

His experience in the US also gave Kato-san the opportunity to act as a bridge between his co-workers in Japan and the team in the US.

> When I was in the US, my co-workers in Japan became my virtual team and it was easier to work with them. One of the reasons I think is that I could help in many situations just as a translator or explaining properly as I could understand both situations.

Although face-to-face meetings are not always essential, in Kato-san's situation, it helped to build a better understanding of his projects, stronger relationships with his team members in the US, and improve communication flows.

> It depends on the situation, but in general face-to-face should help a lot for a virtual team to feel as a real team and sense of unity, also to understand who they are and their personality. It helped a lot to work with them after my return to Japan.

The Internet as a Unifying Language

Communication on the Internet has several advantages, especially when dealing with international challenges. The advantage that e-mail, instant messaging, and other text-based communication tools provide is one that masks an individual's speaking accent. Be it someone from a different country or someone from a different region within a country, text messaging acts as a way to unify a conversation. Internet text messaging also has spawned its own shorthand and jargon. It's common especially in instant messaging conversations to see brb (be right back), btw (by the way), ttfn (ta ta for now), f2f (face-to-face), lol (laughing out loud), rotfl (rolling on the floor laughing). These shorthand abbreviations act as a unifying language across virtual teams. You can find more abbreviations at http://www.netlingo.com/emailsh.cfm.

Chapter XIII

Virtual Team Maturity Curve

Imagine that you live Chicago, and your best friends live in Texas, Florida, California, and Paris. Overall, do you think this is a good thing or a bad thing? Some people might think, "It's a bad thing because I can't see my friends whenever I want, and we can't get together and do things as often as I'd like. We can only talk on the phone and write to each other." Other people might think, "Well, I don't get to see them all the time, but now when I want to go on vacation, I have people in four different places I can visit and maybe even stay with." This is an example of the level of comfort—of preference—that an individual has in a particular situation. For virtual teams, we characterize the level of comfort and readiness to operate in a virtual environment on the "virtual team maturity curve."

It's important to understand the maturity or readiness of a group or individual to work in a virtual team in order to determine which tools and processes will need to be used in order to have them work effectively. For example, we've found that in teams where the members know each other or have worked together before, they are able to get started in a virtual environment more quickly. However, simply having worked together face-to-face in the past is not a guarantee that they will be successful in a virtual environment; there are other factors to success related to tools and processes.

The virtual team maturity curve acts as an indicator of the likelihood of the organization or individual to be successful in a virtual team environment. The curve looks like the following.

The Virtual Maturity Curve

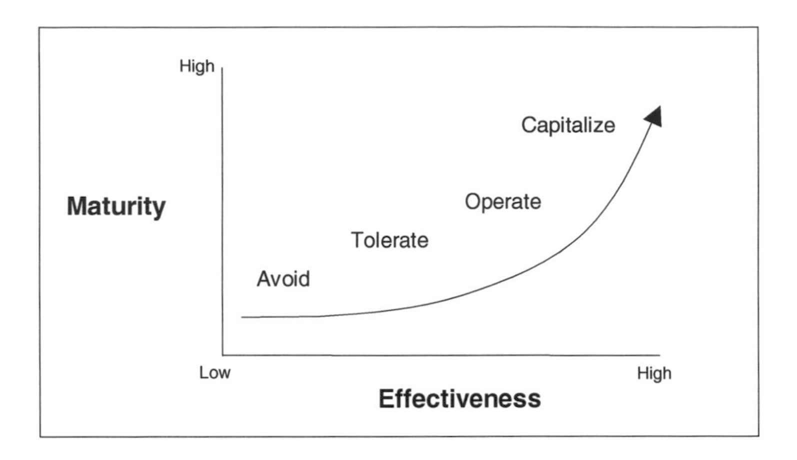

A typical distribution of people in a company or organization looks something like the pie chart "General Distribution within the Maturity Curve".

General Distribution within the Maturity Curve

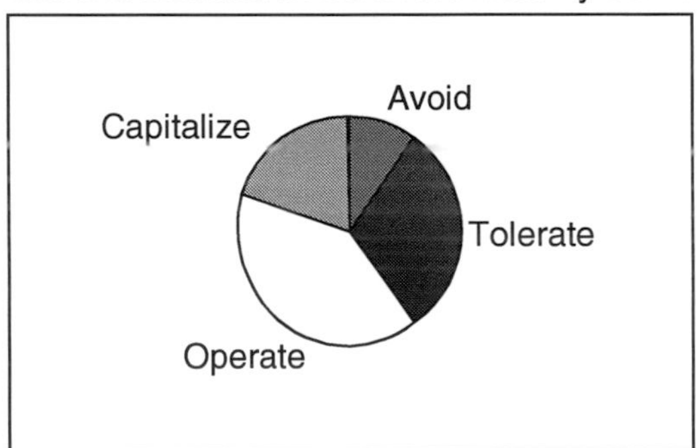

Avoid

Individuals and organizations that are in the Avoid stage typically are uncomfortable or unwilling to participate in a virtual team environment. Typically, in a large population, this stage represents the fewest number of people. People in this stage are unlikely to move up the curve. Individuals in the Avoid stage typically look for projects or careers that allow them to work face-to-face. The interesting trait that we've seen with people in this stage is that geographic distribution is not a barrier to their success. Rather than using tools to operate in a virtual environment, people in this stage tend to have primarily local teams or spend a great deal of time traveling to meet face-to-face. Ensuring that people in this stage spend the majority of their time with a local team or meeting face-to-face will maximize their contribution and satisfaction with their work rather than trying to force them to other parts of the curve. Typically, people in this stage who have supervisory or managerial responsibilities will not be successful at managing a virtual team. There are a few people in this stage who, when exposed to the appropriate tools and best practices of working in a virtual team, comfortably move into the next stage—Tolerate.

Tolerate

People in the Tolerate stage would much rather work in a face-to-face environment, but understand that as a result of the current business environment (i.e., cost restrictions, project requirements, lack of local expertise, etc.), virtual teams are necessary, but they still prefer working in a face-to-face team and believe that's the most effective model. People in the Tolerate stage are flexible, able to adapt to a virtual team environment, and can be very efficient when exposed to virtual team tools and best practices. However, they will not be the ones who will be advocates for virtual teams and may even feel some resentment toward the virtual team model. These are the strongest believers of the virtual team myths. In fact, small minorities of people in this stage are actually in the Avoid stage but forced here out of necessity or directed here by their boss. Typically, one-third of an organization is at the Tolerate stage. If they have enough bad virtual team experiences in the future, they may even move to the left of the curve to the Avoid stage. Here is a list of things to make people in this stage successful:

- Select one or two recommended tools—people in this stage will use tools but won't spend time to figure out which one is best for them.
- Show examples of when a virtual team model has been successful.
- Ensure people understand the benefits.

- Ensure that myths are addressed.
- Make some portion of a project face-to-face, if possible.

When put in a situation where they see virtual teams working, they may move to the next stage—Operate.

Operate

The Operate stage typically represents the largest portion of a population, with over one-third of people at this stage. This group is comfortable with working in a virtual team, probably spends most of their time in virtual teams, and uses some advanced techniques. When given a bad experience, they may move to the Tolerate stage, but more than likely, Operate is where most people remain. They are probably involved in cross-functional teams and possibly involved in a maxtrixed organization with multiple "dotted-line managers" in addition to the manager who is responsible for their salary. This group is open to using new tools and can be a way to introduce new technology to improve effectiveness. For example, in HP, within a period of 12 months, without any formal program, 75% (about 60,000) of the employee population had installed some sort of public instant messaging tool. After four months of a formal program, about 40,000 had installed the enterprise standard IM tool.

Capitalize

People who are in the Capitalize stage are the ones who are actively driving the use of virtual teams in an organization and try to take full advantage of the diversity, cross-functionality, and distributed nature of virtual teams. People in this stage either can be managers who take advantage of virtual teams to give their organization a competitive advantage, or individuals who are evangelists for the virtual team model. There are some people in this stage who feel that face-to-face interaction is not necessary at all or that technology will solve all problems. They may sometimes be frustrated that the tools that they have are not rich enough to perform the advanced tasks that they want to do. They may also be frustrated by people in other stages of the maturity curve who have not embraced the virtual team model as strongly as they have. They are typically the ones who are willing to try out the latest technologies and techniques before they are ready for the mainstream population. People at this stage:

- Can be the catalysts who push the organization forward in the virtual team maturity scale. However, they need to remember often or be reminded that while they pave new paths and move forward, they need to sometimes slow down, or turn around to make sure people are following them and not left-behind.
- Are a source of the most advanced thinking around virtual teams.
- May be a channel for sponsoring or demonstrating new technologies and processes.
- Need to be reminded that the majority of the population is in the Operate stage, not the Capitalize stage.

Organizations and the Virtual Team Maturity Scale

Remember that this scale is not intended to imply that being in the Avoid stage is bad and that the Capitalize stage is good. Rather, it is an indicator of how successful a person or group will be in a virtual team environment. Groups will need to decide for themselves where on the curve they feel they need to be in order for their organization to be successful. Some may decide that because of the nature of their activities, being in the Avoid stage is fine and that moving to any of the other stages is unnecessary. Others may decide that they should move from the Tolerate stage to the Operate stage. In any case, the virtual team maturity scale can help you understand in which stage an individual or group is, what their perspective is on virtual teams, and what their role can be in the overall ecosystem of virtual teams.

Chapter XIV

Where is Your Company or Organization on the Virtual Team Continuum?

Here is a quick quiz to test where your company or organization is regarding virtual team attitudes. We should state up front that this quiz is focused primarily on knowledge worker jobs that don't require attendance at a physical, central location.

You have a team with five people located in a central location and five located in various locations 100+ miles away. When you have an internal staff meeting:

A. The five remote users always have to travel to the central location for a face-to-face meeting (0 points)
B. The five remote users dial in to the central location conference room, where a single speakerphone is used for two-way audio (2 points)
C. The five remote users dial in to the central location conference room, which has microphones for each person in the conference room (4 points)
D. The manager views that since some team members are remote, normal meetings should be conducted with everyone virtual (5 points)

When a new project is announced that will cross organizational lines:

A. The first two weeks of the project are spent deciding which organizational charter owns the different aspects of the project. A common phrase is, "We own that, but we don't have resources to assign to it" (0 points)

B. People are assigned to the project based on their skill sets, not their organizational affiliation (5 points)

During the SARS epidemic, your company:

A. Made no attempt to cut back on travel for internal meetings (as a result of the epidemic) (0 points)

B. Cut back on travel for internal meetings to the most affected areas only (as a result of the epidemic) (3 points)

C. Banned all travel for internal meetings, and went to a strictly virtual meeting model (5 points)

For virtual meetings that involve 50+ participants, your company provides these tools:

A. Telephone bridge only (1 point)

B. Telephone bridge and simple data-sharing tool like NetMeeting (2 points)

C. Telephone bridge and Web-casting tool that allows presenters to flip the slides and have everyone see the presentation (3 points)

D. Telephone bridge, Web-casting tool, and one-way video conferencing (5 points)

Project leaders in your company:

A. Are always managers (0 points)

B. May sometimes be empowered individual contributors (5 points)

When you have a virtual project kickoff meeting with members from multiple organizations:

A. For the first hour of the meeting, everyone that speaks must (through one mechanism or another) be identifiable to the other meeting participants (5 points)
B. People may speak without any means of identifying themselves to other members of the team (0 points)

Performance evaluations are:

A. Always done face-to-face, regardless of the location of the two participants (0 points)
B. May be done virtually, with the consent of both participants (5 points)

During a recession, travel for internal meetings are:

A. Not curtailed in any way (0 points)
B. Approved for customer-facing meetings only (5 points)

To meet federal rush hour environmental regulations, your company:

A. Has the workforce work different shifts, coming into a central location (0 points)
B. Encourages telecommuting so that some employees are never on the road during rush hour (5 points)

In your company, telecommuters:

A. Are assumed not to be working if they aren't at their home desk during working hours (0 points)
B. Are judged based on whether they are meeting their objectives and timelines (5 points)

In your company, use of Instant Messaging (IM):

A. Is banned, because it is considered to be a frivolous waste of time (0 points)
B. Is only allowed between employees using a secure IM system (3 points)
C. Is encouraged, as it is viewed as a valuable tool for geographically dispersed team members to share an electronic "water cooler" experience (5 points)

In your company, most jobs require:

A. The job seeker to move to the location where the job is hosted (0 points)
B. The job seeker to be located anywhere within the country where the job is hosted (4 points)
C. The job seeker to be located anywhere in the world (5 points)

In your company, work is:

A. Something you do (5 points)
B. Somewhere you go (0 points)

In your company, funded IT projects:

A. Are always assigned from the top down (0 points)
B. Can be originated from front-line employees that see a need (5 points)

To see how your company or organization rates, add up the total points you scored from the quiz, and then compare it to the scale labeled "How your company or organization rates".

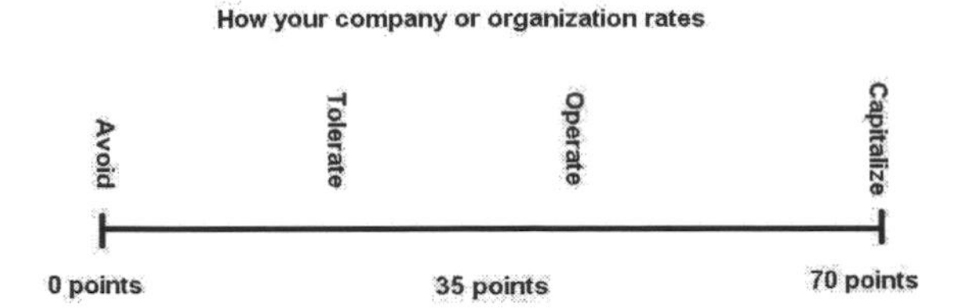

Comments on Our Scoring

You have a team with five people located in a central location and five located in various locations 100+ miles away. When you have an internal staff meeting:

A. The five remote users always have to travel to the central location for a face-to-face meeting (0 points) – *Stone age thinking – "all meetings must be face-to-face."*

B. The five remote users dial in to the central location conference room, where a single speakerphone is used for two-way audio (2 points) – *Rates a few points because the remote participants don't need to travel. However, it is very hard for the remote participants to feel engaged in an environment like this.*

C. The five remote users dial in to the central location conference room, which has microphones for each person in the conference room (4 points) – *The organizer or manager of the meeting is interested in the experience of the remote dial users.*

D. The manager views that since some team members are remote, normal meetings should be conducted with everyone virtual (5 points) – *We assign this the most points because it (1) levels the playing field between the remote and co-located users and (2) recognizes that not all meetings need to have anyone face-to-face.*

When a new project is announced that will cross organizational lines:

A. The first two weeks of the project are spent deciding which organizational charter owns the different aspects of the project. A common phrase is, "We own that, but we don't have resources to assign to it" (0 points) – *A common phrase is, "It is amazing how much time can be lost on critical projects as different organizations jockey for position."*

B. People are assigned to the project based on their skill sets, not their organizational affiliation (5 points) –

During the SARS epidemic, your company:

A. Made no attempt to cut back on travel for internal meetings (as a result of the epidemic) (0 points) – *Meeting face-to-face for internal meetings is worth risking the lives of your employees?*

B. Cut back on travel for internal meetings to the most affected areas only (as a result of the epidemic) (3 points) – *Balances risks versus perceived benefits of traveling under these circumstances.*

C. Banned all travel for internal meetings, and went to a strictly virtual meeting model (5 points) – *Since internal meetings are not revenue-generating, travel for internal meetings is banned to protect employees from any possible risk of infection.*

For virtual meetings that involve 50+ participants, your company provides these tools:

A. Telephone bridge only (1 point) – *Gets a point because the organization is willing to have such a meeting with remote users.*

B. Telephone bridge and simple data-sharing tool like NetMeeting (2 points) – *Two points for trying to improve the remote experience. However, NetMeeting is not very effective for groups larger than 15.*

C. Telephone bridge and Web-casting tool that allows presenters to flip the slides and have everyone see the presentation (3 points) – *A high-quality solution that helps remote users feel immersed in the meeting. These Web-casting tools often have voting and question facilities.*

D. Telephone bridge, Web-casting tool, and one-way video conferencing (5 points) – *Everything the remote user needs to feel immersed in the meeting experience, with a video-feed available to those remote users that have high-bandwidth connections.*

Project leaders in your company:

A. Are always managers (0 points) – *The most advanced project models are often in consulting companies. In those types of companies, one's position on the organization chart usually is not an indication that,* ipso facto, *they are qualified to lead a particular project.*

B. May sometimes be empowered individual contributors (5 points) – *See above comment; in addition, individual contributors often have the highest skill set to lead a particular project, regardless of their rank in the organization.*

When you have a virtual project kickoff meeting with members from multiple organizations:

A. For the first hour of the meeting, everyone that speaks must (through one mechanism or another) be identifiable to the other meeting participants (5 points) – *This is just good audioconference meeting etiquette, where the players don't all know each other.*

B. People may speak without any means of identifying themselves to other members of the team (0 points) – *Very poor audioconference meeting etiquette, when all the players don't know each other. Often can lead to resentment and disengagement of some members of the conference.*

Performance evaluations are:

A. Always done face-to-face, regardless of the location of the two participants (0 points) – *If there are no contentious issues in a performance evaluation, why make the recipient travel to receive it?*

B. May be done virtually, with the consent of both participants (5 points) – *See above comment*

During a recession, travel for internal meetings are:

A. Not curtailed in any way (0 points) – *Your company isn't interested in saving significant dollars in a recession?*

B. Approved for customer-facing meetings only (5 points) – *Travel for revenue opportunities still makes sense, even in a recession. Travel in a recession so that a team can brainstorm its major projects for the next quarter seems frivolous.*

To meet federal rush hour environmental regulations, your company:

A. Has the workforce work different shifts, coming into a central location (0 points) – *Here, the company grudgingly obeys federal rush hour rules, inconveniencing its employees by having some come in at 6:00 a.m. and some work until 7:00 p.m. or later, and so forth.*

B. Encourages telecommuting so that some employees are never on the road during rush hour (5 points) – *Need to cut down on the number of cars your employees put on the road during rush hour? Why not have as many as possible stay at home?*

In your company, telecommuters:

A. Are assumed not to be working if they aren't at their home desk during working hours (0 points) –

B. Are judged based on whether they are meeting their objectives and timelines (5 points) – *a.k.a. Management By Objective.*

In your company, use of Instant Messaging (IM):

A. Is banned, because it is considered to be a frivolous waste of time (0 points) – *Banning Instant Messaging for any reason other than security concerns would seem to be a step backward in the ability for remote personnel to keep in contact with each other and with the central office.*

B. Is only allowed between employees using a secure IM system (3 points) – *In some industries where there is a significant amount of sensitive information being exchanged, using a secure IM system is essential. However, providing that solution only for employees limits extending the virtual team by excluding external partners.*

C. Is encouraged, as it is viewed as a valuable tool for geographically dispersed team members to share an electronic "water cooler" experience (5 points) – *Instant Messaging is a valuable way for remote team members to interact, providing "presence" information and electronically sharing the water cooler experience.*

In your company, most jobs require:

A. The job seeker to move to the location where the job is hosted (0 points) – *This significantly cuts down on the possibilities for your company to hire "the right person for the job, regardless of location." Someone may not want to move to Bayonne, New Jersey, but that person still might be a perfect match for the job you need to fill.*

B. The job seeker to be located anywhere within the country where the job is hosted (4 points) – *Hire the best person for the job, regardless of location.*

C. The job seeker to be located anywhere in the world (5 points) – *See comment above*

In your company, work is:

A. Something you do (5 points) – *This is, of course, a philosophical question. If work is something you do, the location of the work is often irrelevant.*

B. Somewhere you go (0 points) – *If work is somewhere you go, than the emphasis is often on physical presence and not meeting company objectives and timelines ("Joe is always here at his desk, so he must be working").*

In your company, funded IT projects:

A. Are always assigned from the top down (0 points) – *Top down projects can be good because they reflect overall company objectives. However, sometimes innovation comes from the worker bees.*

B. Can be originated from front-line employees that see a need (5 points) – *Front-line employees often have the best ideas about improvements that can be made to company service.*

Chapter XV

Managing a Virtual Team

While participating on a virtual team, a team member has his or her own set of challenges, and managing the virtual team involves some additional responsibilities. A people manager or program manager is expected to lead, be an example for the rest of the team, and help pull the members together to accomplish something collectively that they could not accomplish alone. Time, budget, and scope are the three main attributes of a program that can be documented, managed, and measured. Operating principles and culture are the soft skills that can contribute to the success or failure of the program or project on which a team is working.

Be an Example

If you are managing a virtual team, you will personally need to be in or quickly move toward the operate or capitalize stage of the virtual team maturity curve. Leaders of a virtual team who are in the avoid stage are bound for failure, while those in the tolerate stage will not enable the team to be as effective as it could be. Whether you are aware of it or not, your team members will be looking to you to set the tone for how the team will operate. Even if you do not consider yourself an expert at managing virtual teams, you will be more effective if you understand the common myths and best practices, and, more importantly, if you can leverage the diversity of knowledge and expertise within your team. Find people who are in the capitalize stage to help you build your repertoire or tools and best practices. The goal of leading a virtual team should be to ensure that the team feels like they are working, not being on a virtual team and trying to get their work done. The fact that they are on a virtual team should not be a distraction or limitation to accomplishing their overall objectives.

Best practices for managers:

- Be an example
- Get off to a great start
- Standardize
- Be visible

Get Off to a Great Start

Here are some ways to make a virtual team project launch successful:

- Ensure you spend enough up-front time in planning.
- If you need virtual breakout rooms, look for those types of capabilities in audioconferencing bridges and set up separate real-time application sharing sessions.
- Make sure you keep to the agenda or build in enough flexibility so that agenda items can be moved or be shortened.
- Ensure that there are action items assigned and that follow-up discussions are scheduled to reinforce the point that this is a working meeting, not just a discussion, and that the real work will come later.

Standardize

Deciding which tools and best practices you will use up front will help avoid confusion and simplify the way your team operates. The following are the major decisions you will need to make:

- How will you store your project materials? Will you use a document repository, a shared file folder, a shared directory on a network, a Web site, or perhaps just e-mail?
- What tools will you use to communicate within the team? Is e-mail fast enough? Do people check their voicemail? Is instant messaging too intrusive?
- How often and at what time will you meet? If the team has members in different time zones, you need to determine if there is a realistic common time that everyone can get together. In a team with many different time zones, sometimes breaking up the team in subgroups based on time zones is a feasible solution; other times, we've seen a "share-in-the-suffering" model where each of the major time zones takes turns participating in meetings at odd hours.

Time Zone Considerations

When building a virtual team, you may find it critical to have team members from different time zones participate. However, the difficulty in managing the team grows as the number of hours between team members grows. It's almost a given to work with people across time zones with one to three hours difference. When there is an eight- to 10-hour difference, the situation is still manageable, but typically, someone will be working early in the morning or late in the evening. When the time difference is greater, it is extremely difficult to coordinate meetings and collaborate in real-time.

Here are a few suggestions to minimize the pain of working across time zones:

- Ask people what time zone they are in and refer to a world clock or world time service such as http://www.timeanddate.com/ or http://www.worldtimeserver.com/ before setting up a meeting. By doing so, you can avoid scheduling meetings during other people's lunch hours, dinner times, or in the middle of their night.
- Just because the majority of the team is near a particular time zone, don't ignore the one or two people who may need to wake up at 5:00 a.m. or stay at work until 7:00 p.m. so they can join the rest of the team. Try to

find a mutually agreeable time so that everyone feels like an equal member of the team.

- If you must create a team with members across Europe, Asia, and the Americas, try to minimize the time when the entire team needs to get together, and try to divide the work in such a way that people near each other's time zones can work together.

Keeping these suggestions in mind when building a team and scheduling events will help to build trust between the team members (or at least help avoid resentment) and ensure that all team members are concentrating on the topic to be discussed rather than on the meal, sleep, or personal commitment that they are missing.

Be Visible

One of the major challenges in managing a virtual team is ensuring that there is visibility inside the team in relation to what each member is doing, and that outside the team, sponsors and stakeholders are kept informed of status, issues, and milestones.

Inside the team, it will be essential to have regular status meetings and checkpoints so everyone understands their responsibilities and that dependence on other team members and their deliverables are clear. Publishing an overall plan of record, or dashboard, to track these items is one way to accomplish this task. These documents also could be leveraged for communicating outside of the team.

Outside of the team, announcing a program or project launch can help make the team and project visible and may create opportunities for links for similar or follow-up projects. It also may help prevent redundant efforts and identify related activities that may be leveraged. Once the project is completed, it is extremely important to publish your results and celebrate the completion. Team members may feel that reward and recognition for their efforts may not be as high when they are in a virtual team because they may not have the opportunity to be as visible. It's up to you as the manager or program manager to ensure that the team, its accomplishments, and all its hard work are visible and that the team contributes to the satisfaction of each of its members. Later in this book, we'll discuss ways to celebrate your accomplishments in a virtual environment.

> Unless the manager and the team have worked together before, the manager of a virtual team needs to spend a dedicated amount of time early on to build trust and credibility.

A Note for People Managers

There is a mysterious effect when a person changes his or her role from an individual contributor to a people manager. People who once saw this person as a peer now view him or her as an outsider. What was once a casual conversation or an off-the-cuff remark now becomes a statement of direction, a requirement, or perhaps a cloaked indicator of some hidden agenda. This has the same effect when a manager is put in charge of a virtual team.

Unless the manager and the team have worked together before, the manager of a virtual team needs to spend a dedicated amount of time early on to build trust and credibility. In the past, if you were a manager of a new team, you would gather the group together in a room so that everyone could meet you, shake hands, and get to know you. When some or many of the members are distributed, getting together in a room is no longer a simple matter. Travel costs as well as time considerations are major factors. The speed of business only seems to accelerate, and there are not many organizations that have the luxury of several months of startup time. In fact, in some cases, a team is expected to execute as soon as a team is formed.

One of the authors has spent 10 years managing remote employees. During that time, there have been two instances where he's never met an employee face-to-face, and there was no time that he felt a remote employee had less of his time or attention than a local employee. In fact, in his experience, he spent more time with remote employees since they made a conscious effort to get together. Since he assumed he could get together with local employees at any time, he actually ended up spending less time with them. Here are a few key things that will ensure you are managing a remote employee successfully:

- Establish a rapport; get to know the person first.
- Make yourself available—if they don't call you, give them a call or send a quick instant message just to check-in to see how things are going and if there's anything you can do to help.
- Schedule regular one-on-one time.

- Come prepared with topics to discuss.
- Everyone is different, so each meeting will be different (frequency, types of topics, etc.).
- Find out what interests the person—useful for development as well as rewards and recognition.
- Establish clear goals and measures; review them periodically.
- Be flexible and sensitive of time zones.
- Find out what motivates them.
- Make sure their contributions are visible within the team, to management, and outside the team.
- Don't just talk to them when you want them to do something. Ask them for their opinions, advice, and perspectives.

Rewards System

As a manager, you can change your team's behavior by rewarding certain actions. This is true for encouraging the use of virtual teams and propagating best practices. Publicly recognizing an individual or team for running an especially successful virtual event, or complimenting someone for the way they have contributed to a virtual team encourages more of the same behavior. Publishing and distributing best practices to the organization or group aids in moving people up the virtual team maturity curve. And by highlighting the person or team that initiated the best practice, you also are recognizing them for positive behavior.

However, great caution should be used to avoid the interpretation that participation in a virtual team is itself a reward. For example, telling someone they can work at home occasionally should not be used as a reward for high performance. This only propagates the misconception that being part of a virtual team is a perk rather than a standard operating model.

Chapter XVI

Managing a Virtual Organization

As the business environment becomes more dynamic and virtual teams become more prevalent, organizations cease being physical groupings of people and transform into functional groups that are spread geographically.

The virtual organization, like the virtual team, takes advantage of distributed expert knowledge and cost efficiencies. But like managing a virtual team, managing a virtual organization presents some challenges. Town hall meetings, coffee talks, organizational announcements, beer bashes, celebrations, and motivational events all require clever alternatives to the traditional face-to-face events.

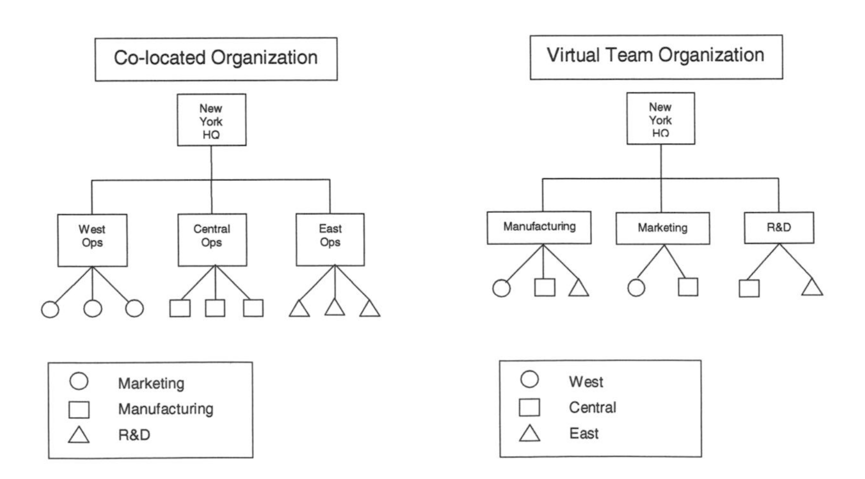

Announcements

In a co-located organization, getting everyone together for an important announcement is challenging, but it is still a realistic option. When companies were just starting to form virtual teams, we saw important announcement events scheduled in such a way that all the face-to-face teams in different cities and states heard the same message at exactly the same time. The manager of a large organization might send one of his or her local staff members to San Francisco to host a face-to-face meeting on Wednesday at 1:00 p.m., while another staff member goes to Boston to host a face-to-face meeting on Wednesday at 4:00 p.m. It is important that everyone hear the same message at the same time to prevent the East Coasters from getting the information during the morning before the West Coast has had a chance to hear the information from their assigned executive. Today, it's more common to handle this situation through Web casts and audiobridges.

Town hall meetings or coffee talk sessions with the organization have also moved from face-to-face to virtual venues. We had seen, over time, the audience for face-to-face coffee talks shrink on a regular basis as a reflection of the distributed nature of the team. Soon after that, you lose critical mass; it's somewhat embarrassing to have a general manager travel to a site and have 10 people in the room and 150 people on the phone.

Motivation

Another challenge in managing a virtual organization is how to keep it motivated. In later chapters, we'll talk about virtual team building and celebrations that can be used to address this challenge. The other major hurdle for the leadership team is accessibility to the employees and visa-versa. A few best practices we've seen work in the past include the following:

- Communicate frequently – if the employees cannot see you frequently, ensure that they hear from you so they are kept up-to-date with organizational priorities, challenges, and direction.
- Be frank but sensitive – especially when facial expressions are not visible and the only emphasis is in the tone of your voice, it's important that your

messages are clear and straightforward in order to build credibility, but not so blunt that the message is misinterpreted.

- Be accessible – make yourself available via instant messaging and town hall meetings; participate in project reviews.

These suggestions not only build credibility for the management team, but they also act as a strong way to pull together the virtual organization to ensure it continues to be focused and informed.

Setting Up Your Organization for Success

The design of each virtual team, supporting tools and processes, and training and development all contribute to make an organization successful overall in utilizing virtual teams. Many virtual teams are created out of necessity because of cost or timing constraints. But without any planning, virtual teams can be very ineffective.

The design of each virtual team should be considered carefully. It would be a mistake to confuse representation with participation. Having representation in a virtual team might mean that each major geography (e.g., Europe, Asia, the Americas) or each major business function (e.g., Marketing, R&D, Sales, Finance, etc.) needs to send someone to be a member of the team to represent their specific needs. On the other hand, participation in a virtual team is driven not by ensuring that every population is represented, but rather that the key skill sets and knowledge that are critical to accomplishing the team's goal are included in the team. Sometimes, as in the PC COE example described in Section Two, representation is a critical success factor, but it is not necessary for all virtual teams.

Supporting tools and processes enable the virtual team to be successful. In Section Four, we'll describe some best practices for taking advantage of tools that support the operation of a virtual team as well as how to perform certain face-to-face activities in a virtual team setting. One of the most important considerations is to try to standardize on as few tools as possible. This will reduce the overall cost to the organization as well as help propagate best practices and shorten the learning curve, since everyone will be using the same toolset. For example, Hewlett-Packard has an internal portal called "Collaboration Central" that lists recommended collaboration tools and best practices. This portal is available to all HP employees and fosters the effective use of these tools across the company and with its business partners.

Training and development are the final components needed to help an organization effectively operate its virtual teams. Most organizations spend most of their resources on obtaining and deploying tools without proper training. Most people don't bother to attend training because they are too busy. However, we've found that most people are able to learn how to use a tool effectively once they've participated in a well run virtual collaboration session. The following guidelines can help an organization provide high impact training and skills development at minimal cost.

- Some rudimentary training should be made available either directly through the tool's help facility, through self-paced training, or through the broad publication of best practices, tips, and tricks.
- Focused in-depth training should be provided to project managers and team leaders who will be leading virtual teams and who will have the best opportunities to act as role models for how to best take advantage of tools.
- The amount of training you invest in should be proportional to the impact that the tools can have on the overall effectiveness of the virtual team—not the overall cost of the tool. Just because a tool is low cost or potentially free doesn't mean there should be no training provided. This is especially important when it comes to security. Although there are many free instant messaging tools available through the Internet, most of them are unsecured and transport your conversations in plain text. Without proper training, an organization could lose sensitive information to anyone who might be trying to monitor this type of network traffic.

With some up-front planning, an organization can increase the effectiveness and success of virtual teams though the design of virtual teams, deployment of supporting tools and processes, and the availability of training and development opportunities.

Section IV

Best Practices for Virtual Teams

In this section, we will discuss best practices for virtual teams. We also will talk about the tools available for use by virtual teams as well as some of the techniques we've seen employed for team building within virtual teams. We'll include a brief discussion of situations where virtual meetings might be preferable to face-to-face interactions.

Chapter XVII

Impact of Culture, Affiliation, and Shared Goals

As we mentioned in Section 1, technology provides only a portion of the solution for addressing the challenges of virtual teams. Team and organizational cultures are key indicators of the potential success of virtual teams, while shared goals provide a way to unify the group. Team affiliation, as we will see, plays a lesser part in ensuring the success of virtual teams but is a reality of the complex organizational structures we see today and evolving in the future.

Team Culture

In this discussion, we expand our definition of a team to include small groups of two to three people working together to thousands of people working in a large organization. In these team environments, the team leader or sponsor will determine the culture of the team. Sometimes this is called the team's values, beliefs, or operating principles. How the team will work together is stated either explicitly in the culture or implicitly by the way teams or team members are rewarded. Sometimes, what is stated explicitly contradicts what happens in reality. For example, a company could say that it values diversity, but its hiring practices could imply that diversity is valued only to the extent that it doesn't interfere with the homogeneity of the organization.

A virtual team can be successful only if the team leader or sponsor supports a virtual team model. In our experience, when there was strong sponsorship for working collaboratively in a virtual team environment, the team was most successful. Even if the team members are enthusiastic about a virtual team model, without strong sponsorship, the rewards and recognition system and lack of funding for tools discourage the continued operation in a virtual team environment.

Companies operate under many different models, but general preferences can be extrapolated by looking at where companies like to hire people. In January 2004, we went to the job Web sites of three global high-tech companies: HP, IBM, and Microsoft. We compared the total number of job openings with the number of job openings in the state where the company's corporate office was located. At HP, 44% of the job openings were located in the same state as its corporate headquarters. At IBM, that percentage was 34%; at Microsoft, it was 82%. We compared this with similar data from years past and saw a similar pattern. Although this short survey was far from scientific, this type of analysis can give you an indication as to whether a company relies mostly on co-located teams or distributed teams in its day-to-day operations.

	% of jobs in state (Mar 2002)	% of jobs in state (Jan 2004)
HP	46	44
IBM	21	34
Microsoft	87	82

Virtual Out of Necessity

In some cases, teams are virtual out of necessity. Typically, small teams with limited resources or specialized knowledge need to rely on expertise and knowledge from other groups in order to accomplish their goals. The scientific and academic research community is an excellent example of how virtual teams can flourish. Experts in highly specialized fields must collaborate and share knowledge in order to further their research.

How Many Different Hats Can You Wear?

As a result of complex lifestyles and ever changing matrixed organizations with multiple dotted and solid reporting lines on organizational charts, team and individual affiliations are having less effect on the success of virtual teams. In the past, people worked in a single team as part of a single company, and their affiliation was clear. But that clarity is now a long lost luxury of the past. Today, people have multiple affiliations—they wear multiple hats. In your private life, you may act as a mother, wife, basketball coach, and first-string violinist. Similarly, in a virtual team, it's likely that you have multiple affiliations, and emphasizing any particular affiliation won't necessarily make the virtual team more successful. The more important thing to focus on is ensuring that the team has a shared goal, objective, or purpose.

Shared Goals Strengthen Virtual Teams

Once you have adequate tools and a culture that supports virtual teams, the last critical ingredient for success is identifying a shared goal. We've all participated in teams where the objectives weren't clear, we weren't sure why we were even getting together, or we had general ambivalence toward the subject matter. Some people make the distinction that a "group" is just a collection of people, while a "team" is a group of people that need to rely upon each other to obtain a shared goal. We've seen examples of groups of people that were brought together because they had similar roles in the same geographic location. The members of the group operated independently of one another, and each had its own individual goal. There were overall goals for the group, but they were not dependent upon each of the team members working together.

The most successful teams we've managed and participated in all had a shared goal and required each team member to work together in order to achieve the goal. Without the shared goal, the virtual team is no more than a collection of people grouped in some manner.

The Top Leaders Set the Tone

Ultimately, it is the topmost leader that determines the overall culture of the organization and makes a conscious or unconscious decision to operate at a specific point on the virtual team maturity curve. Like the team leader or sponsor who sets the tone for each individual team, the topmost leader has the largest impact on the success or failure of all virtual teams in the organization.

If the policies of the organization and the behavior of its topmost leaders are in the avoid stage of the virtual team maturity curve, and they encourage only face-to-face interaction, then it will be difficult to get funding or attention from them to move up the curve. Likewise, if the leaders are at the most extreme part of the capitalize stage, they may question why anyone should travel to meet face-to-face. The hiring policies and day-to-day operations revolve around the preferences of the leaders. If the leaders are at one extreme or the other, it is almost impossible to move their organizations toward the center of the curve. However, most of the organizations we've interacted with are either in the tolerate or operate stages. In these cases, there are some ways to evolve the culture of the organization and move it up the virtual team maturity curve.

Changing the Culture

While individual trust, tools, and training are important in moving a person or team up the virtual team maturity curve, there is a totally different set of approaches to changing an organization's culture and moving the entire organization further up the curve. Logic may steer you toward the idea that a culture is built by the beliefs and behaviors of individuals. Therefore, in order to change the overall culture, you start at the individual level. While that may be true when looking at social culture, an organizational culture is strongly influenced by its topmost leaders. Therefore, an organizational culture change needs to start from the top.

If the leaders and their organizations are in the tolerate stage of the curve, here are a few things that can be done to shift their mindset and, as a result, evolve the culture to be more receptive to a virtual team operating model. The main objective is to show that the virtual team model can be a competitive advantage to the organization.

- Demonstrate the business case for moving up the virtual team maturity curve. We've discussed the business case in great detail earlier in Section Two.
- Highlight success stories and show their impact on the overall organization. More difficult but even better would be a comparison between the results achieved through the virtual team and estimated results if the team was required to be co-located (e.g., increase in cost, increase in time to completion, ideas that would have been missed).
- Give examples of what competing organizations are doing and how their use or lack of use of virtual teams is affecting their overall performance.
- Outline the steps that can be implemented to help individuals move up the virtual team maturity curve.

If the leaders and their organizations are in the operate stage of the curve, then they already understand the importance of virtual teams, but they need to be shown how their organization could be more effective.

- Illustrate costs that could be reduced or avoided by simplifying and standardizing on a smaller set of recommended toolsets.
- Demonstrate how best practices can increase the effectiveness of virtual teams by reducing the time it takes to complete a project or by increasing the richness of the final result.
- Paint a long-term roadmap of what could be accomplished in the future by moving the organization up the virtual team maturity curve.

Once the leaders see the benefits of moving up the curve and they feel that movement is important for the success of the organization, then the culture of the organization can start to change, and the individuals will find support in their own personal movement up the virtual team maturity curve.

Chapter XVIII

Evolution of the Groupware Concept

In the early 1990s, Lotus created the concept of groupware with its revolutionary (for the time) product, Lotus Notes. For the first time, Lotus Notes tied together into one package various components needed for collaboration (an especially attractive premise for the SMB marketplace)—directories, electronic mail, group calendaring, file sharing, chat rooms, and so forth. This was the first attempt at creating a collaboration environment for teams and organizations to foster a sense of community and improve information sharing and communication. Early versions of Lotus Notes tended to be very top down—a central administrator was needed to set up a chat room, file sharing area, and the like. While Lotus Notes wasn't very flexible, it was (in a pre-Web, pre-Intranet world) an attempt to improve collaboration at a time when teams were becoming increasingly geographically and organizationally dispersed.

Over time, market changes have evolved the concept of groupware, sometimes pressuring it to centralize (e.g., Microsoft Exchange) and sometimes pressuring it to devolve into "best of breed" products. The Web provided a new framework for groupware and made it easier to include business partners (outside the firewall) to participate in shared collaboration areas. Improvements such as better document review mechanisms, more flexible administration, and more granular security also followed.

One interesting philosophical change in groupware (by the late 1990s, it was better referred to as "teamware" or "team collaboration") was Groove, a product developed by Lotus Notes creator Ray Ozzie. The basic version of

Groove is peer-to-peer based, and, unlike earlier groupware products, requires no IT administrative overhead (Groove also offers a version that allows IT oversight). Groove also can work through a firewall, allowing easy integration with business partners and customers. Philosophically, the base-Groove is perfect for small virtual teams that need to quickly create a shared area for calendaring, file sharing, chat, and so forth.

Most companies, however, want to balance flexibility in a team collaboration project, with some degree of IT control. Products such as eRoom and Windows Sharepoint Services try to strike a balance between the two. While still maintaining ultimate IT control over the system, a team, once it has been granted access to a shared space, has great flexibility to create spaces for document sharing, calendaring, chat, and so forth within the shared space.

Two recent Web developments push the limits of user control of the collaboration process—blogs and Wiki. According to the "Blogphiles," Web ring, blogs are defined as follows:

> A Web site (or section of a Web site) where users can post a chronological, up-to-date e-journal entry of their thoughts. Each post usually contains a Web link. Basically, it is an open forum communication tool that, depending on the Web site, is either very individualistic or performs a crucial function for a company (http://www.blogphiles.com/webring.shtml).

Blogs have been utilized to provide on-the-spot commentary from international hot spots (i.e., a lone Iraqi providing daily updates on the progress of the US invasion in 2003) to jump-starting a political campaign for President (i.e., Howard Dean's campaign). By combining wireless technology with blogging, people can provide up-to-the-minute commentary on unfolding events such as political conventions, stockholder meetings, or natural disasters.

Blogging tools allow even novice Web users to post their thoughts on the Web without understanding arcane Web technologies, letting them concentrate on the *content* and the not the *form* of the content. Some commentators believe that blogging is a real revolution (as opposed to evolution) in the use of the Web, as it helps make the Web a two-way participation medium, rather than a one-way, content-delivery mechanism. Everyone can be a publisher.

Stretching the boundaries of self-administered team collaboration tools is Wiki, which is described as "a collection of Web pages which can be edited by anyone, at any time, from anywhere" (http://c2.com/cgi/wiki?WikiGettingStartedFaq). In this case, policing the site for things like obscene language, inflammatory remarks, and the like is not controlled by restricting access (the traditional groupware approach), but rather by devolving the policing function to the broader community.

Whether centrally administered or given decentralized control, groupware products have come a long way since their creation in the early 1990s. They remain a key element in the ability of geographically and organizationally dispersed teams to communicate, innovate, and share.

Chapter XIX

Applying Tools for Maximum Impact

Early on, we mentioned that technology won't solve all problems related to virtual teams. In fact, basic meeting management and project management skills and techniques are actually more important than trying to use the most sophisticated tools. For example, if you don't have a meeting agenda with specific topics to discuss, identified decisions that need to be made, and owners for action items, the best virtual meeting tool is not going to automatically do those things for you. At the same time, if we didn't have the tools we have today, virtual teams would not be possible, and the speed of business would be driven by how many face-to-face meetings we could squeeze into our schedules. Identifying and using the best tools for each situation is a key to maximizing the effectiveness of every virtual team. In many cases, people often start by selecting a tool, try to use it for every situation, and end up frustrated or confused. This is not surprising, since it's like going to the kitchen, pulling out a bag of flour, and thinking, "Now what can I do with this?" rather than deciding you want to bake some bread and what you really need is bread flour. Let's look first at what we want to do and then identify the most appropriate tool for that task.

We've grouped tasks into four major categories. This was not done after many years of scientific research; there are as many ways to categorize tasks as there are tools to perform those tasks. We came up with these groupings based on the most common things people want to do in a virtual team.

Conversations – Conversations can be real-time live conversations either through voice or text. They can involve two or more people, but usually, if you have more than four people involved, three to four people at most will participate actively in the conversation. In some situations, speed of information exchange is most important, while at other times, ease of joining the conversation is most important.

Document Storage – In the physical world, document storage would be in filing cabinets or libraries. In the virtual world, you still need a place to store your documents and make them available to the people who need to have access to them and protect them from those who should not. These people may be inside or outside of your team, department, organization, company, subsets, or combinations of them all. Secure storage and easy access to authorized team members is the most important aspect here.

Real-Time Sharing – Real-time sharing involves visually sharing information that is either static (e.g., slide presentations for a small group or large classroom audience) or dynamic (e.g., real-time editing of a document). This activity can involve two or 2,000 people effectively. Usually, these interactions include an associated phone call or conference call. The main requirements here are some type of network access and secure or even encrypted connectivity.

Presence – Presence is not really a task or activity, but it is probably one of the most overlooked, powerful, and unappreciated elements of an effective virtual team. Typically a feature of instant messaging tools, this function lets people on your contact list know whether you are available, on the phone, busy, or away. It is a major tool for helping physically distributed teams feel closer by allowing each other to "see" what they are doing. The most important thing here is selecting a single common platform or a platform that is compatible with multiple platforms.

Let's Talk

Conversations are at the heart of getting work done in a virtual team. People are spending a large amount of time, sometimes the majority of their day, either processing e-mail or talking on the phone. Instant messaging is quickly replacing quick phone calls or e-mail, although many people are still exchanging four to five quick e-mail replies in a matter of minutes as another form of instant messaging. But at the same time, we are getting overwhelmed with the amount of data we receive and the number of physical and virtual meetings we need to attend. How can we effectively hold these conversations?

A few things to consider when selecting a tool for conversations:

Is *speed of interaction* important? Do you need responses immediately, or can you wait a few hours or days for a reply?
Is *archivability* of the conversation important? Do you need to save the conversation for future reference or for legal reasons?
Is *speed of setup* important? Do you need to have the conversation immediately, or do you have time to set something up? Typically, audioconferences need to be reserved.
Is *richness of interaction* important? Is the tone of voice or tone of the message going to be critical to avoiding misunderstandings?
Is *accessibility* important? Will people be in a place where there is network and computer access, or will they be on the street with a mobile phone?

The following table identifies the most effective tool based on the requirements of the conversation.

	Speed of Interaction	Archivability	Speed of Setup	Richness of Interaction	Accessibility
E-mail	O	▲	O	O	O
Phone call	▲	▽	▲	▲	▲
Instant messaging	▲	▲	O	O	O
Text messaging	▽	▲	▲	▽	▲
Video messaging	▽	▽	▽	▲	▽
Voicemail/ voice messaging	▽	▲	▲	▲	▲
Audioconference	▲	O	O	▲	▲
Discussion Forums/ Newsgroups/Threaded Discussions	O	▲	O	O	O

▲ = high
O = average
▽ = low

Here are a few best practices to keep in mind:

Audio Conversations

A typical audioconference goes something like this: imagine that you are Lee, a project manager. Tom, a project manager for another group, has asked you to join his team meeting to help them solve a problem. You get on the phone conference just as it begins. This is what you hear:

Tom: Ok, let's start the meeting. Is everyone here? Terry? Sam? Mary? Linda? Lee?

Male voice #1: Yes

Mail voice #2 and a woman's voice together at once: Yes

You: Yes I'm here.

Tom: Ok. I asked Lee to join today to give us some advice.

Male voice #2: I think it's too early for advice. We still have a few things to try.

Female voice: Were those tests run yesterday?

Background noise: We are now boarding flight.... row numbers.... have ID...

Male voice #1 (or was that Male voice #2?): Yes, both were negative.

Tom: What was that?

Male voice #1: I said, "Both were negative."

Tom: Ok, we better look at the test results. I'll start NetMeeting.

Male voice #2: I won't be able to join. I'm at the airport waiting for my flight.

Tom: Oh, that's what that noise is.

Female voice #2: Hi, sorry I'm late.

Tom: No problem, who just joined? Lee's here. We were just about to ask for some advice.

Later in the meeting:

Female voice #1: Those are great suggestions, Lee. We'll try them later today.

Male voice #2: I couldn't hear everything. I'll read the notes after we land. Who's taking notes?

Tom: I've got a few. I'll send them to you.

You: I've got some too I can send.

Female voice #2: Great, send me a copy too.

You: Sorry, who was that?

Female voice #2: Linda.

Certainly, this is not the best way to have an audioconference. How could this conversation be improved?

- Perform a roll call and introductions at the start of each call so the participants know who is on the line and what their roles are.
- If you are in a group conversation with people you don't speak to on a regular basis, make sure you identify yourself each time you speak so people know who is talking. "This is Mabel, let's think about...."
- Typically when the main speaker on a phone conference asks the group a question, there will be total silence, which is usually followed by, "Is anyone still there?" Usually, it's more effective asking one or two people for their input by name.
- Mute your phone in noisy places. If you are in a noisy environment, it will be hard for you to hear. But if you don't mute your phone, you're sharing that noise with everyone on the phone. Of course, remember to unmute the phone before you speak.
- From an audio conversation, you may wish to jump quickly into some type of real-time application sharing. Before you do that, think about the tools to which people have access. Some may be sitting at a desk and have network connectivity, while others may be driving in their car on a mobile phone. If you know you will need to share materials, make sure they are available before the meeting.
- Be cognizant of the other person's native language. If participants are not fluent in the language in which you are going to have your conversation, and if you are planning on having a complex conversation, speak slowly, check for understanding, or perhaps send e-mail instead.

Let's try the same meeting with the best practices:

Tom: Ok, let's start the meeting. First we'll go around the phone and do some quick introductions. Terry?

Male voice #1: I'm Terry, and I'm the test engineer.

Tom: Terry, can you be the note taker?

Male voice #1: Sure, no problem.

Tom: Sam?

Male voice #2: I'm Sam, and I'm the lab manager. I'm at the airport waiting for a flight. It's a bit noisy here, so I'll mute the phone.

Tom: Mary?

Female voice: I'm Mary, and I'm the design engineer.

Tom: Linda?

(Pause)

Tom: Looks like she might be late. Linda is another lab engineer who just wanted to listen in. We can start now, and she can catch up through the notes. Lee?

You: I'm Lee, and I'm a program manager from the other lab. We've seen similar problems to those you're experiencing.

Tom: Ok. I asked Lee to join today to give us some advice.

Male voice #2: This is Sam. I think it's too early for advice. We still have a few things to try.

Female voice: This is Mary. Were those tests run yesterday?

Male voice #1 (or was that Male voice #2?): This is Terry. Yes, both were negative.

Tom: Ok, we better look at the test results. I'll start NetMeeting.

Male voice #2: This is Sam. I won't be able to join, but I've got hard copies of the file you sent with the meeting request.

Female voice #2: Hi. This is Linda. Sorry I'm late.

Tom: No problem, Lee's here. We were just about to ask for some advice.

Later in the meeting:

Female voice #1: Those are great suggestions, Lee. We'll try them later today.

Male voice #2: This is Sam. I couldn't hear everything. I'll read the notes after we land.

Male voice #1: Ok, this is Terry. I'll send you a copy as soon as we hang up.

Female voice #2: Great. This is Linda. Send me a copy, too.

> Other tip: Some audioconferencing services provide breakout rooms so you easily can have subconferences spawn off the main conference. This is useful when holding a large main meeting. Divide the group into subteams, have them go off on their own conference, then return to report back to the large group.

Text Conversations

- Unlike voice conversations, it's easy to have multiple instant messaging conversations going on at the same time. It's a great opportunity to multitask, but you should set the other person's expectations as to how fast you're able to reply.
- Before sending someone an instant message (IM) out of the blue, first ask them if they are available to IM. The last thing someone needs while they are projecting their computer screen in front of the room during a presentation is an IM window flashing up with a private message.
- Tagging the subject of your e-mail can help the recipient manage and prioritize messages. Starting the subject line with words like "ACTION," "FYI," "URGENT," or "FEEDBACK" can help people manage the huge volumes of e-mail they receive. Having groups standardize on a set of common tags is useful but typically difficult to implement if people don't decide to use them all the time.
- Sending Web links or URLs to people instead of attaching files helps to prevent filling up your recipient's mailbox and causing frustrating delays if they are downloading your message through a phone line. For most recipients, this is useful, but for people who usually read their mail disconnected from the network, they won't have access to the data.

Video Conversations

- Typically, video conversations are good for an initial formal meeting where people have not met before. Because of the overhead in setting up video and the limitations in network bandwidth, video between groups that know each other or will only be working together for a short period of time is not worth the overhead.
- Probably the best use of video is sending prerecorded video clips. We've seen excellent uses for training, major announcements from executives, or just short greetings.

Virtual Filing Cabinet

Storing a document, presentation, image, or other piece of electronic data is useful only if you can find it again. Sounds pretty basic, but without any planning, you could easily end up with data spread out across different storage systems, embedded in various applications, or more likely filed away somewhere on your e-mail server. In a virtual team, it's important to share information with people who need access and to protect it from people who shouldn't have access. For those people to whom you allow access, it should be easy for them to find and access the content, make modifications, or add new content. For those people who are restricted, they shouldn't even know the data exists.

When considering how to share information teams should think about:

- Audience
- Access
- Authentication
- Aging and Archiving
- Arrangement
- Amount

The main things to consider when you're trying to figure out how to share information with your team are:

- **Audience** – Do you need to share information with a small workgroup of three to five people or a larger extended team of 10 to 20 people? If you want to share something with more than 20 people, you probably want to think about ways to publish the information broadly rather than just how to store it for a team.
- **Access** – How will the information be accessed? Does the information need to be available through devices like PDAs and phones? Will the person need to be able to download files for off-line access to the information? Is it possible to have everyone install a specialized client software application, or is a Web browser the only thing people will have? Will the data need to be encrypted as it is transmitted?
- **Authorization** – How closely do you need to manage access? Do you need to create access lists per individual file, or all files in a particular folder?
- **Aging and Archiving** – Is it important to be able to identify all files older than a certain date and move them out of the virtual filing cabinet onto tapes or disk for long-term storage?
- **Arrangement** – Probably one of the most important decisions you will need to make: Should I group information by project, by type of activity, or by organization?
- **Amount** – How much data do you need to share? Ten files or 1,000? Small 1KB files or large 10MB files?

Show and Tell

Real-time application sharing tools and services like NetMeeting and WebEx are usually used to share slides and other documents in small groups for review, editing, and discussion; or in large groups, primarily in classroom style presentation and training.

But there are less common uses of these tools that virtual teams can take advantage of.

Brainstorming

Most people have participated in group brainstorming sessions. A common one involves affinity diagrams (or a similar process called the KJ method, developed by Japanese anthropologist Jiro Kawakita). The activity goes something like this: Everyone gathers in a room. Index cards or sticky notes are handed out to each participant. Everyone is given five to 10 minutes to jot down ideas with the following guidelines: don't filter any ideas, think big, generate as many ideas as you can, and so forth. When time is up, everyone shares their ideas and posts them on the wall. People start to group them arbitrarily, based on common themes they identify in the ideas, and create labels to describe the group. The wall usually ends up looking something like this:

Group 1	Group 2	Group 3	Group 4	Group 5
Idea	Idea	Idea	Idea	Idea
Idea	Idea	Idea	Idea	Idea
Idea	Idea	Idea	Idea	Idea

This resulting list is called an affinity diagram. Sometimes the major groups then are grouped into two or three major themes.

This process easily can be performed in a virtual team using the shared whiteboard functionality of a real-time application sharing tool. The shared whiteboard allows all participants to post their input to the whiteboard simultaneously. Creating an affinity diagram in a virtual team goes something like the following.

Everyone dials into an audioconference. They then launch a shared whiteboard. Everyone is given five to 10 minutes to start typing ideas onto random areas of the whiteboard. Often, participants will select a color for themselves to use for typing so the idea can be associated with the participant later in the discussion in case clarification of the idea is needed. If the input is to be anonymous, then everyone uses the same color. Once time is up, people are able to click on each of the ideas and move them around the whiteboard to group them. Discussion can happen via the audioconference, and resulting categories can be assigned.

The major side benefit of the electronic affinity diagram exercise is that in the end, there is an electronic record of the resulting groupings and ideas that can be shared easily and distributed, avoiding the face-to-face problem of assigning a person to carefully collect all the cards from the wall and painfully transcribing each of the cards, many of which contain quickly scribbled fragments of text.

We've used this technique and seen this technique used effectively for small groups of five to 10 people up to large groups of 50 people.

Voting

We're not talking about electronic presidential elections here, but rather a commonly unused feature of real-time conferencing tools that allows the facilitator to create electronic polls and surveys. These polls can be used in several ways.

- **Testing for understanding** – Remember when we discussed the awkward silence during an audioconference when the discussion leader asked the rest of the group a question, and the leader is rewarded with pristine silence? Inserting polls two to three times during a presentation to test for understanding or ask for input ensures that the audience can participate actively in the discussion.
- **Anonymous input** – Typically, the polling feature allows anonymous input. This may encourage people to express their true opinions on controversial or sensitive issues.
- **Evaluations and feedback** – Polls can be created quickly to get quick feedback about the presentation (or online class) and its effectiveness at the end of an event. In fact, they are so easy to create on the fly that one of us was making a one-hour presentation when halfway through the event, we forgot to set up an evaluation for the end. We quickly sent an instant message to one of the other presenters and asked them to set up the poll while we continued to present. Five minutes later, we received an instant message saying it was done.

Markup

In high school, it was painful to sit in English class and have your teacher hand back the five-page essay that took you three weeks to write and see it marked up with red ink on every inch of each page. There's only one thing more painful than that: to watch your document being marked up in a real-time conferencing tool by your entire virtual team.

The idea seems logical enough—group document markup in real time to quickly revise documents and receive all input at the same time. In practice, this activity is neither quick nor efficient. In a group editing session, typically several people will have different and often conflicting feedback. Discussing and evaluating all that input at the same time is overwhelming, and consensus is never reached.

A more effective way to gain input on a document from your virtual team is to have each of them annotate a copy of the document, send it back to you for review, and then have a final review with the team to show them which changes you've incorporated and which changes you've rejected.

Another method involves a real-time markup session with no more than one or two people and a clearly identified leader who can make final editing decisions.

Are You There?

One of the most common problems when working on a virtual team is the feeling of isolation. Even if you are in a building with other people, they may be people you occasionally work with or, more likely, a random face that decided to use the open-seating cubicle next to yours for the day. If you're a manager of a virtual team, you need a way to see quickly which one of your team members is around that can quickly answer a customer question, help you deal with a sudden emergency, or take your place in a meeting at the last minute. In a co-located team, you could walk down the hall, pop your head up and look around the surrounding cubicles, or check the cafeteria. In a virtual team, you need to rely on presence.

Most people view instant messaging tools as a way to send quick text messages to chat to people, and very few have the discipline to remember to change their status from "online/available" to "on the phone" and back again after they compete their phone call. But that short status information, or presence, is critical for letting your virtual team members know whether you can be interrupted, or if you are away from the computer. Many document storage applications include presence indicators, while real-time conferencing applications include mood indicators or the ability to virtually raise your hand.

Here are a few habits to form:

Be conscious of your presence and always change it to reflect your current status.
If you can customize your status message or name, include other relevant details (e.g., meeting@alaska room, out of office, @customer site, etc.).
If you can customize your name, don't change it to just your initials, first name, or something cute. Seeing "KP," "gr8guy," or four "Bobs" on your IM contact list is not helpful.

Additionally, most IM clients allow you to set up a private chat room or IM session with multiple people. This is very useful for inviting two or three people to a quick ad hoc discussion to make decisions.

Sharing Information in Virtual Teams

There are many types of information that must be shared among team members, whether they are virtual or co-located teams. However, information sharing can provide challenges for virtual teams due to the fact that methodologies often employed by co-located teams are not available in the virtual world. Examples might include:

- **Routine project documentation and updates** – In the co-located world, these types of updates might be provided through face-to-face project meetings, posters on bulletin boards in the coffee room, and so forth (e.g., come to Conference Room A today at 3:00 to hear about the progress of the XYZ project).
- **Day-to-day organizational announcements and bulletins (i.e., updates, status, etc.)** – In a co-located team, these types of announcements might be posted on bulletin boards, distributed via hard copy to in-trays on the desk, and so forth (e.g., we are rolling out a new tool for scheduling conference rooms; new security access procedures will be in place next Monday, etc.).
- **Organizational announcements that contain significant new information (e.g., reorganizations, etc.)** – This category would include announcements that have the potential to be very controversial or that may

generate a lot of questions. In many cases, for co-located teams, these types of announcements might be made through all-hands meetings where team or organization members gather in a large conference room, cafeteria, and the like to hear the message.

There are certainly different profiles for the intended audiences for information, including:

- **One-on-one** – This would include informal information sharing among individuals.
- **Small groups** (15 or less) – This might typically characterize information sharing among a team profile, whether it is an organizationally based team or a project team.
- **Large groups** - A large group could be a hundred people in an organization, the entire company, and so forth.

Of course, the tools that can be applied for information sharing can vary, based on the type of information and the intended audience. Here are some methodologies (and the associated tools) that we've seen used to address information sharing needs within virtual teams.

- Routine project documentation and updates
    - Project Web sites can be utilized to communicate key information about project status, including project plans, frequently asked question, project contacts, and so forth.
    - Periodic e-mails communicating status – E-mails can be used to effectively distribute routine status reports regarding project or team activities. In many cases, e-mail distribution lists can be subscriber based, where users can sign up to receive updates on projects that interest them.
    - Project teams can hold periodic audioconference meetings, inviting anyone interested in hearing a project update. These audioconferences can be supplemented by meeting management tools (such as NetMeeting) for sharing presentations, if needed.

- Day-to-day organizational announcements (updates, statuses, etc.)
  - These types of routine announcements typically can be completed through e-mail notification, as they tend to be less controversial in nature. Specific e-mail messages can be used to address an individual topic, or several topics could be handled through regularly published, organizational newsletters.
  - A pointer to a Web site on the company intranet (either an organizational or company portal, depending on the nature of the announcement) can be included to guide readers to where more information is available (e.g., FAQs, ways to get more information, etc.).
  - If they are not time-sensitive in nature (i.e., the word has to get out today or this week), these types of announcements are sometimes handled through periodic (i.e., quarterly, etc.) all-hands meetings that are conducted utilizing audioconferencing and Web casting tools.
  - Organizational announcements that contain significant new information (e.g., reorganizations, etc.) – In a virtual team or organization, these types of announcements are often addressed through multiple mechanisms.

Possibilities include:

- All-hands meetings combining audioconferencing for those who can connect only via telephone, as well as an accompanying Web cast for those with high bandwidth connections, can be utilized for these types of announcements. These types of meetings also are often recorded for off-line playback by individuals who were unable to attend.
- A follow-up e-mail (using organizational or company-wide distribution lists) is sometimes utilized to summarize the announcements.
- Readers can be directed to a Web site on the company intranet (either a company or organizational portal, depending on the nature of the announcement), where people can get more information, access FAQs, or ask a question.

We'll summarize these observations on sharing information within the virtual team environment in the following table.

	One-on-One	Small Groups (<15 people)	Large Groups (>15 people)
Routine project documentation or updates	▪ Project Web sites	▪ Project Web sites ▪ Audioconferences (possibly combined with meeting management software) to discuss project status and answer questions	▪ Project Web sites ▪ Audioconferences (possibly combined with meeting management software) to discuss project status and answer questions
Day-to-day organizational announcements and bulletins	▪ E-mail updates ▪ Phone conversations	▪ E-mail updates ▪ Organizational newsletters ▪ Company or organizational portal	▪ E-mail updates ▪ Organizational newsletters ▪ Company or organizational portal ▪ All-hands meetings held via audioconferencing / meeting management software
Organizational announcements with considerable new or controversial data	▪ Phone call	▪ Audioconference combined with meeting management software ▪ Follow-up via company or organizational portal ▪ Follow-up via e-mail summary	▪ Audioconference combined with Web cast. ▪ Follow-up via e-mail notification ▪ Follow-up via company or organization portal

Summary: Some tools available for sharing information on virtual teams:

- Project Web sites
- Audioconferences
- Phone conversations
- Meeting management software
- E-mail updates
- Organizational newsletters
- Company portal
- All-hands meetings

Chapter XX

Team Building

Team building as a whole is generally looked at as a non-essential or a waste of time. In most teams, it is just ignored. In others, varying degrees of effort are made to build a sense of camaraderie. These range from exchanging digital photographs of each other to playing interactive games.

Photographs

One of the easiest ways to pull a team closer is to have them exchange digital photographs of each other. Associating a face to a voice on the phone helps individuals build an image of the person they are dealing with and makes the person real. Ways that we've seen photographs enhance the virtual experience include the following:

- **Visual organization charts** – Rather than publishing a document containing names in boxes connected by lines on a chart, we've created and seen charts that replace each of the boxes with photographs. This is most helpful for team members and people working with the team, but we've found that it adds almost no value when shared with a general audience who is not working directly with the team.

- **Web meetings and documents** – Some Web meeting tools allow you to post a picture of the speaker in the corner of the Web client while a viewer is watching slides. Similarly, we've seen document authors include their pictures either at the beginning or end of a paper that they've written. Some members of our teams regularly include their photographs as part of their e-mail signature. Each of these techniques helps to build a more personal connection between the speaker/author and viewer/reader.

One major advantage in the use of digital photographs is its relatively low complexity compared to live video. Almost any digital camera can produce a high enough resolution picture that can be easily transferred to most computers and then incorporated into a document. Video, on the other hand, is a bit trickier to ensure a high quality image when considering the varying quality of cameras, importance of lighting, compatibility with various operating systems, and computer power requirements. One team member, Carol Wolf, once said, "What people expect is Hollywood-quality video with high fidelity sound, professional lighting, and multiple camera angles. Instead, they get a poorly lit, fixed camera, with low resolution and jittery movement."

Sometimes, Face-to-Face Doesn't Matter

We've seen that when you've used digital photographs effectively, and you have been working with someone over a long period of time on a regular basis, meeting with them face-to-face doesn't add much to the interaction. In fact, there were several face-to-face meetings that we've attended where it was actually anti-climatic to see people that we regularly work with in our virtual team. We spent most of our time talking to those people that we interacted with less frequently.

Virtual Pictionary

During one staff meeting, we were trying to think of ways to hold a virtual team building event, and someone came back with the idea of "Virtual Pictionary."

We wanted an event that we could hold virtually that would be fun, that used the tools that people were already familiar with, and that didn't require us to buy anything. On the morning of the event, our entire lab of 20 people, distributed across the United States and Europe, dialed in to a conference bridge and launched NetMeeting. One team member would draw, while the rest of the team had to guess what it was they were drawing. We would take turns so that everyone had a chance to show off their artistic skills (or lack thereof). So that no one else could see, the host used instant messaging to send the current "artist" the thing they needed to draw. These were things like a sunset, fishing, and so forth. We would open a whiteboard so that everyone could see, and the person would then start drawing on the whiteboard while everyone tried to guess what it was they were trying to represent. Since we had a global team, after a few rounds, someone blurted out how to say the word in Finnish. Others added translations for Chinese, German, and French. So, for the rest of the game, part of the fun was guessing the drawing and then translating it into as many languages as the team knew. It turned out to be a terrific way to have a team-building event.

As a result of using virtual technology to hold this event, people could attend from whereever they happened to be. One of the participants was sitting in his children's elementary school parking lot, having just dropped them off for the day. Using a Tablet PC and wireless connection to the Internet, he was able to participate as easily as everyone else in the event.

Everquest

In one of the organizations that we worked in, we held innovation sessions that brought people together from across the organization into a forum where they could present and share ideas for areas of research. One idea that originated from those sessions was to determine how to use virtual reality environments as a tool for team building.

The team wanted to do something more complex than hold a one-hour event to play Virtual Pictionary. They wanted to create an entire virtual environment where a small group met to interact. They decided to explore massively-multi-player-online-role-playing games (MMORP) and used a game called Everquest as the virtual environment. Two groups of four to six people were selected to interact in Everquest a couple times a week, for about one hour at a time, for a period of three months. This was done either during lunch hour or after work when typical informal team building happens in a face-to-face environment.

In Everquest, each player selects an avatar, or computer representation of themselves, as their character. The game has a medieval theme with swords to swing, monsters to hunt, and treasures to find. Unlike many games, this one emphasized teamwork and so provided an ideal environment for experimenting with virtual team building.

At the end of the three months, we concluded that although the environment was successful at building a stronger team, the technical overhead and time commitment was much too high to implement as a formal way to build a strong virtual team. Some of our findings were as follows:

- The virtual environment gave team members a chance to get to know each other better and understand their personal styles based on how they reacted in certain situations. Some people rushed in to attack monsters, some wanted to take some time to plan, while others stayed in the background and healed other characters when they were hurt.
- The environment required a certain level of computing power that most people didn't have at the time. This required separate equipment that was loaned to them specifically for the purposes of the experiment.
- The game provided a two-dimensional image of the environment from a first-person point of view (e.g., the images you saw represented what you were seeing as if you were looking through the eyes of your avatar). This caused some people difficulty since there were no physical cues to help them judge distance, and many people got lost while they were exploring the environment. Others experienced motion sickness as their avatar ran through the virtual landscape.

In a virtual team, just like on any other team, building camaraderie is important, especially if the team will be working together on a regular basis. Since you can't go out and have lunch with each other, you need to find other ways to socialize, even if it means just spending the first or last few minutes of a phone conference chit-chatting.

Chapter XXI

The Challenges of Virtual Teams

We wanted to get a perspective on the challenges of virtual teams from people who prefer working face to face rather than in a virtual setting. Greg Todd has had roles as a manager and an engineer on virtual teams. Joe Gerardi moved from Georgia to California to manage a team based on the West Coast.

Joe comments in general about virtual teams as well as the challenges around meetings and people management. He also shares his views on how culture and tools can act as an enabler or barrier to the success of virtual teams. He considers himself about three-fourths of the way into the tolerate stage of the virtual team maturity curve.

Continuums

Joe: If you ask me if face to face is better than virtual, there isn't a single answer. You need to consider several continuums: Is the content highly structured or variable? Is the content local only or does it apply globally? The physical distance of team members can then vary from people living together to being totally remote.

If you build a see-saw and put on one side the overall value (overall productivity and cost) of having a virtual team in the long-run and having a physical team that you can pull together infrequently, the virtual team wins out. But for specific situations, the newer the thing is, face to face wins out. If it's new and we haven't done it before and it requires creative thinking and needs full engagement, you need face to face. If you are already up and running, you can get as much out of the remote team.

Meetings

Joe: If you plan the meeting and have an agenda, you can get a lot of work done. But there are many challenges and barriers in setting up virtual meetings. Time zone differences are a major barrier. Folks in Europe are amiable in accommodating schedules in the US. You always feel you're imposing. Sometimes it feels more fair to spread the aggravation around by rotating meeting times.

People seem to do a lot of email during virtual meetings. Using instant messaging for side chats relevant to the discussion topic is an appropriate thing to do during meetings. But sometimes people are just chatting about something totally unrelated.

Also, you get barking dogs, dishes, and crying kids in the background. People just don't mute their phones. Cell phones are not really suited for phone conferences because the transmission quality is still not there.

In a virtual meeting, you need to explicitly check for consensus. You don't get body language clues. The visual input is not the key thing. It's the overall physical interaction that's important. There is conscious and unconscious feedback when you are face to face. When you're virtual, you only get the conscious feedback.

Virtual is not as good as being in the same room at the same time. People are reading body language. You are aware of disagreement sooner. You can tell if people are actually engaged. You can see people and what they are doing. When you're there you're there. Certain processes like kick-off kinds of things and brainstorming kinds of things have higher fidelity in person.

I haven't seen a situation where everyone was remote that has generated a better end product compared to when everyone was face to face. You can have more meetings with a larger variety of people remote but not as high fidelity.

People Management

Joe: For remote management, you mainly follow good standard manager rules. For example, you need to treat everyone individually. Some employees like to talk a lot, some like to talk a little.

I wouldn't fire someone remotely.

Performance evaluations should be done face to face. Remote is ok, but face to face is better.

I would not hire someone without meeting them face to face. I would interview them remotely. But you can't make a hire decision until you meet them face to face. However, I could make a "no hire" decision without meeting them face to face. It also depends on the type of job. Swapping tapes in a data center is different than doing strategy. You need to ask yourself, what is the level of risk? How unique is the set of required skills?

Culture

Joe: For a virtual team to be successful, you need to have people who are open to it. People need to be willing to try working in an environment where you can't easily go have lunch together. Sometimes there are cultural preferences around people being in offices and cubicles. In those situations its common to hear someone say, "Go find Jim and bring him here now."

If you're working with a culture that doesn't use virtual teams very much and you end up having half of the people in a room and the other half on the phone, homogenizing by making everyone be on the phone helps.

With virtual teams you have a big challenge around "esprit de corps" since you just talk on the phone. When I managed a distributed team, the folks in California really enjoyed it when they got together every four weeks face to face. If you ask them, they might not be able to say why, but they enjoyed it.

If you make a decision to rely heavily on virtual teams, you also need to invest in some face to face. Most of us are gregarious (biologically speaking).

Tools

Joe: Technology is not much of a challenge anymore. Instant messaging helps with ad hoc communications. The tools are getting better. Technology is getting better. It's still up in the air if video helps. Video doesn't help if you don't know the other person because the quality is not good. If you do know them, then it's not essential, it's superfluous. Viewing the object being discussed is more important than viewing the people who are speaking.

The Argument Against Virtual Meetings

If you're going to have a distributed workforce, you've got to sign up for people to travel, and go and meet each other periodically. It doesn't have to be every other week, or even every month, but it should be once a quarter. You need to get out to see people or have people come to see you, whatever it is, and having that personal interaction. (Greg Todd, Hewlett-Packard)

Regardless of how good virtual meeting tools get, there are people that just don't like meeting virtually, feeling that too much is lost by not being able to see people as you interact with them. Greg articulates situations where he believes face-to-face is better.

Greg: Team meetings are much better in person because you can see people's faces, you can play off the non-verbals in the room, you can much more easily address unspoken concerns. People tend to be free to talk more when they're in person, because they feel more a part of the team, they feel much more of a connection there.... It's also much more efficient because you can feed off each other—you can see hand motions, you can see if somebody wants to draw on the white board. You can physically see and interact with each other. So, if you are trying to tackle a complex topic, it is much easier to do that in person.

When you're talking about people's career, or people's personal problems. If you're trying to do someone's performance review—it's really, really awkward to do on the phone. (Sometimes it's not too fun to do in person, either!) But it goes much better in person. If I'm a manager, I'd rather tell someone good news or bad news in person. If I'm on the receiving end as an individual contributor, I'd rather hear it straight from that person, and be able to look them in the eye and talk about it.

Let's say you're up for a promotion and you don't make it. It's much easier to explain what happened in person than it is over the phone. Over the phone, you're missing a lot of the communication. For example, if I'm telling someone, "here's why you didn't get the position that you wanted," I immediately want to see the look on their face, I want to see the look in their eye, and how they react. If they cry, or say, "yea! I'm happy not to do that," I want to be able to see that. But a lot of that is nonverbal, and you can't catch it on the phone.

Conversely, **if you're giving someone good news** – "you just got a raise" or "you just got some stock options" or "you're going to get to do this project that you really wanted to do," it's great to see people's face light up with excitement, to see them enjoy what they're doing and get into it, and just experience that good moment with them. If you're on the phone, you're not really able to experience that with them, and they miss out on the benefit of experiencing it with you. Making that connection with people is really important.

I can't stand doing **interviews** on the phone. A phone screen is fine, but that's only to screen them. But if I'm really going to interview them, it's got to be in person.... You need to see how they interact personally, how they interact with other people, watch how they carry themselves, and how they conduct themselves. Are they sweating or not, are they nervous or calm? You get a sense of person just by the personal interaction. And sometimes what happens when I'm doing an interview is you "click" with the person. Sometimes you're with someone and it just doesn't click. How that person interacts, clicks, and is able to adapt, and is able to work with me, and with other people that are doing the interviewing is something that is going to be critical in someone's ability to perform. Let's face it, unless they're locked in a room cranking code, and someone slides them food under the door, people are going to have to interact with people. So you have to be able to test that during an interview. Although it's somewhat artificial because it's an interview cycle, at least you get a look at it.

Doing **brainstorming and creative work** – it's impossible to do on the phone. It's something I've tried it before, and it doesn't work. I've done a lot of brainstorming sessions, and you have to have people present. I can remember one example where we were doing some brainstorming, and we had some people in the room, and we had some people on the phone. And the people on the phone just ended up dropping off after a little while because they said they couldn't see what was going on, they couldn't see the interaction. We would stop and have side conversations, and people would draw things on the board and say "see this, this is what I mean," and the poor people on the phone had no clue what was going on. And they just miss out on a huge amount.

Resolving conflict. Let's say you've got a couple of engineers who've both come up with a really great technical solution or something. You can do that on the phone, you can do it in a conference call, but if there is emotion attached to it, or if there is maybe some personal agenda attached to it, or if there is some other "thing" going on there aside from the objective evaluation of which is the better solution, then it is much easier to handle that in person, because you can quickly see that that is what is going on. And you can handle it much more effectively.

Section V

Moving Forward with Virtual Teams

In this section, we will look ahead to the future, highlighting how the use of virtual teams within corporations may evolve in the future as well as how future generations may influence the use of virtual teams. We'll also present a narrative that highlights how virtual teams may meet in the 2010 time frame, and present a quiz designed to help determine where your company or organization is on the virtual team continuum. Finally, we'll end with a quick look at the question, "Do virtual teams *ever* need to meet face-to-face?"

Chapter XXII

The Virtual Organization

While this book has highlighted the successful (and some of the less than successful) techniques and applications of virtual teams, the future evolution of virtual teams is yet to be seen within the marketplace. Clearly, many external forces within the marketplace could potentially either drive further utilization of virtual teams within organizations, while other factors might point toward a diminished use of such models. In this chapter, we'll discuss three potential paths that the use of virtual teams within the enterprise might take, and discuss in more detail the scenario that we feel is most likely.

Possibility Number One: Virtual Teams Fade in Prominence

In this scenario, the use of virtual teams would eventually fade within corporations. Their use would essentially go the way of other management and business fads, with the pendulum swinging back to utilization of more co-located teams and organizations. The remote worker becomes the exception to the rule and must try to fit in as best they can. Factors that might influence this evolution might include a desire of the return to more face-to-face communications and team methodologies, based on feedback from team members that might begin to feel too isolated. Corporations may also determine that the business benefits from the use of virtual teams do not outweigh the challenges.

Possibility Number Two: Virtual Teams Remain in Use, But in More of a "Niche" Role

In this situation, virtual teams would continue to be used within large corporations where they made business sense; however, they would not necessarily become a pervasive way for corporations to conduct all business. The majority of business interactions would be conducted within face-to-face, collocated teams. This situation may arise because technology advances stall, making more broad-scale implementation of virtual teams unfeasible, or because corporations discover that in some cases virtual teams just don't make sense. Basically, organizations would be asked to make virtual teams work, where necessary, while putting a primary focus on face-to-face teams.

Possibility Number Three: The Virtual Organization Emerges

This possibility describes the evolution that we believe is in the future for virtual teams within corporations. In this scenario, virtual teams are taken to the next level, and the operations of the entire corporation embrace the virtual working environment. Employees are geographically distributed in many locations, and the ability to work in this environment is an expected skill set for all workers, including knowledge workers, sales associates, support staff, factory personnel, middle management, development engineers, and even the executives.

Much like the Web links disparate computing systems together, corporations would become a web of employees, all linked through technology, that work together to solve business problems and bring products and services to market, regardless of their geographic location or team affiliation. Teams are more fluid, engaging to solve business problems and disbanding when no longer necessary, as employees move on to the next challenge. The key business driver would be bringing the right person to the job or task, regardless of location.

In a completely virtual organizational model, a company would potentially no longer have a home office or central location of any sort. The executive staff is also geographically distributed, with key executives in the areas most needed to run business operations. This type of model has been featured and discussed in industry press, such as *Business Week*, where such operations were labeled as "transnationals" ("Borders are so 20th Century," *Business Week*, September 22, 2003).

In our view, the virtual organization trend will be fueled by the dynamic that the business problems that geographically distributed teams can help solve (as discussed in Section 2) will not go away permanently over time. As the world becomes more connected, as markets continue to move at a breakneck speed, and as global competition increases, companies constantly will be challenged to do things better, faster, and cheaper.

Although worldwide economic conditions surely will improve over time, there will always be another wave when conditions are less than favorable. As the business world becomes more global in operations, profit margins are falling and competition is increasing. Companies that are well positioned to save money, regardless of the economy, clearly will be at the advantage. The need to move more quickly and adapt to new marketplace demands will continue, as well. Teams that can quickly come online and disband, as needed, will enable companies to bring new products and innovations to market more quickly. Providing business continuity in the wake of unforeseen world events will also continue to be a goal for corporations. Corporations that are well positioned to continue operations, despite the inability to travel to or keep offices open in selected locations around the globe, are clearly at an advantage. Finally, the need to leverage skilled workers, regardless of their geographical location, will continue. We feel that the days of mass-scale employee relocations (and their associated expenses) are diminishing. For reasons already discussed, corporations no longer can afford the time and energy involved in relocating new hires (as well as employees that join a company due to mergers) to a central hub.

Moreover, the technology that can help link geographically distributed organizations will continue to improve, providing an improved infrastructure to support the operations of the virtual corporations. Examples include improved video-conferencing, more pervasive and faster broadband connections, and the like.

The virtual organization may also include interactions with business partners and other players within a company's value chain. Increasingly, there is a trend toward businesses focusing on core competencies and outsourcing the rest. This brings new business partners to the mix that must be integrated into core business functions. And, when selecting business partners, corporations are going to want the flexibility to select the best company for the job—not the company that is located in the same city, same geographic location, and so forth. Once again, the web of the virtual organization expands.

The marketplace is also seeing an increase in experimentation with bringing shareholders, yet another set of corporate players, into the mix electronically. Companies are beginning to experiment with virtual shareholder meetings—allowing shareholders to participate remotely via Web conferencing tools.

Finally, the impact of younger generations, as they move into the workforce, also may play a role in this evolution and should not be under estimated. As future generations are more comfortable with technology and non-face-to-face interactions, they will impact the evolution of virtual teams. Generations that grow up communicating with far-flung friends via tools such as e-mail, instant messaging, participating in distance learning courses where they may never meet their professors face-to-face, and so forth, are sure to impact the acceptance of more virtual ways of working.

Virtual Team Case Study: Eddy Current Specialists, Inc.

In Section 2 ("Business case") we introduced a company named Eddy Current Specialists (ECS) of Villa Rica, Georgia, for their use of geographically dispersed regional representatives among their analysts. However, the company has taken the model a step beyond being simply geographically dispersed. While the company has analysts scattered around several states, it only has two actual employees—both office workers in Atlanta. The technicians are all subcontractors who, at the end of the year, share in the profits generated by the company. In an interview with company president Ken Eisenhauer, we learned the advantages and disadvantages of this e-lancing model. On the advantages side of the ledger (from both the subcontractor and corporate standpoints):

Analysts have a sense of personal ownership for doing a good job for their customers. They're not just on the clock.

> There are other companies out there that have employees for analysts. That employee makes X dollars per year and may get some kind of bonus. But it doesn't make any difference whether he does 50 or 100 machines; he still has his base salary. He's got a paycheck continually coming in.
>
> It doesn't work like that for us. If we don't work, we don't eat. As far as testing goes, our level of accuracy and concern for doing the right job—a great job—is at a much higher level than he [an employee] is, because no matter if he screws up, he's still on salary. If we screw up, we may never go back. And therefore, we'd lose customers. So it behooves us to do a super job, to do it right, to be courteous, etc., etc., because if we don't we may not be back again.

ECS only has two people on the payroll, which provides administrative advantages at tax time. Analysts own their own equipment, buy their own medical insurance, and procure their own drug testing certificates—none of these things are ECS's responsibilities.

> They get a 1099 every year—they do their own taxes. They're not on our payroll. We have two people on our payroll, and they're the administrative people. They handle the paperwork and the scheduling. We're all subcontractors, so we're liable for our own workman's compensation—it's not an office concern. They're liable for their own drug testing, they're liable for anything a subcontractor would be liable for, including liability, their equipment, etc. It's not an expense that the office has to incur.

An average analyst makes two to three times more than an employee in the same business typically makes. Analysts share in the company profits at the end of the year.

> Another plus is the analyst makes more, seeing as he "owns the company," at the end of the year, any of the profits work out to be profit

> sharing, and it goes back to him. Whereas an employee, he's going to get what he's going to get, and the guy that owns the whole thing is going to have a fat wallet. Everything here is pumped back to the people that do the work. We probably make on an average two, and sometimes three times what an employee would be making, doing the same thing.

All analysts have a direct impact on company decisions—because, essentially, the analysts are the business.

> If you take a traditional company structure where you have the company at the top, and you picture the lines going to the bottom to all the employees—an organization chart—for us, turn it upside down. Essentially, this company is designed for the workers. If you want to look at it, it's actually a co-op. All these people as a conglomerate work together and say how the company is going to go.

Ken also listed some of the disadvantages of the virtual organizational model of ECS:

Taking disciplinary action can be difficult, as all of the analysts are subcontractors. For example, it is sometimes difficult to ensure timely submission of reports to the central office.

> Getting reports in on time can be a problem, because he's [the analyst's] not here. Reports have to be assembled, and it probably takes on average an hour to assemble it. Ultimately, I can't make them send it in on time.
>
> If they're an employee, I could take immediate disciplinary action. But how do you take immediate disciplinary action against an equal? So, sometimes there is a bit of a conflict there. But for the most part, no.

Since decision-making can involve all of the analysts, it is sometimes difficult to agree on a clear direction, and analysts often want to do things their way.

> The down side of this is, with everyone being subcontractors, every once in a while they want to assert their independence. Technically, we're all equal. Sometimes, when I ask for opinions on a direction we should take, I get 10 different opinions from 10 different analysts.

Finally, Ken notes that it was challenging to find an insurance company that understood ECS's virtual organizational model ("You want insurance on subcontractors?"), but they were eventually able to.

While ECS doesn't consider itself in the vanguard of the virtual team movement, it has created a very efficient virtual organizational model that is very cost effective.

Chapter XXIII

A Virtual Ecosystem

Just as e-business began as a specialized and separate operating and delivery model in companies and has since become a standard way of doing business, virtual teams, organizations, and business models will be the way that things are done. Once we reach a fully virtual business ecosystem, we see two major roles emerging: (1) highly specialized, efficient, and optimized content providers, and (2) customer focused, value-added integrators.

In an optimal virtual team model, each participant brings specialized and unique knowledge, perspectives, and skill sets. The diversity of the group creates a richness in idea generation and a tendency to avoid group think since the team comes together for a fixed period of time for a specific project or objective—there is not enough time for people to homogenate.

Because team members are dependent on each other to contribute knowledge that they do not have, one of the most critical roles to be played on the team is that of the program manager or integrator. The integrator needs to focus on the overall solution, depending upon the team members to provide each component of the solution. The integrator also must identify those things to customize or add in order to optimize the final solution. Thus, in a virtual ecosystem, like a virtual team, the integrator has the most important part of pulling each of the individual parts together and forming the final solution.

In a business ecosystem, this virtualization is not limited to parts and products but also applies to services. Many companies have virtual human resources (HR) organizations where the company HR department no longer provides the services, but rather acts as the integrator to bring various external health care services together to provide a complete health plan for employees. The process of making a movie is very similar. Usually, it is the producer that acts as the integrator, bringing together specialists (e.g., writers, directors, actors, costume artists, editors, etc.) in a virtual team to create a film.

Success in a Virtual Ecosystem

A few things to keep in mind:

- You don't need to do everything yourself. Leverage expert knowledge.
- Results can be better and richer through the diversity of a virtual team rather than relying upon what you can do yourself or who happens to be co-located with you.
- Time and costs can be reduced through virtual team models by taking the approach of divide and conquer.
- Decide what role you will play in each situation, the role of the integrator or content provider.

Chapter XXIV

E-Lancing

Forming teams in a virtual environment is typically an ad hoc process. Usually, project managers draw from talent inside their organizations or from people that they or other team members know from previous interactions. You're really limited by who you know or by the size of your extended social network. So how do you know that you have formed the strongest team possible with people who have the specific knowledge and experience that your project needs? Ideally, as a project manager, you could advertise the types of skills you are looking for, and people could apply to work on that project based on their availability, expertise, and interest.

Working with MIT

In August 2001, we launched a research project with MIT to experiment with a new way to staff projects. Tom Malone and Rob Laubacher had been working on a concept called e-Lancing where freelance workers and employers come together in the open market through the Internet to form virtual working teams. We wanted to take that concept and move it from an open market environment into the enterprise to determine if it would be an effective model for staffing technology research projects.

We were members of a 100-person team distributed across 70 different locations focused on strategy, research, and development. The organization was formed around the idea that teams should not be static and hierarchical but rather fluid groups that were gathered together based on common interests and skill sets. Howard Bain was the leader of this organization and described why this model was used rather than a traditional organizational structure. "We don't want to build silos. We feel most breakthroughs come about when you mix and match people with different kinds of expertise."

Supply and Demand

Typically, when staffing a project, you have a supply of people resources and a demand from the projects. In our alternative staffing model, we wanted to introduce the concept of creating a supply of projects and managing the demand for those projects by team members. After creating a set of potential project ideas, we advertised the ideas and resources requirements through e-mail.

Sometimes, project teams would hold peer reviews in which individuals could come together in a forum and review an idea or project and provide input to its creation or design.

Members from the organization could see what new projects were available and express interest based on their availability and skills. Using this model of advertising available projects and allowing the entire organization to view and select activities to work on had the following advantages:

- Staffing was not limited to people who were known to the project manager or other team members—projects were exposed to the entire organization. People who previously were not considered for a project because they were not the obvious choice now had an opportunity to contribute.
- Team members had more control over how they spent their time, since they could select the projects they wanted to work on rather than be given an assignment.

We also created a database that contained every employee in the organization. Employees could edit their own data to include a list of their technical knowledge, skills, and interests. So, if an employee was too busy to read or happened to miss an advertised project, a project manager could search the database for potential team members.

Projects were now staffed with people who were passionate about the subject, had the strongest skill sets, and leveraged the diversity of team members who might not necessarily be technical experts in a particular area but brought in different perspectives. In the end, the project results were richer, and the employees gained more satisfaction from the work.

Additionally, there were a couple of beneficial, but unexpected side effects, of this model. Because some of the project advertisements ended up being forwarded to other employees in the company, we started to attract interest from people outside of the direct organization and were able to start drawing upon an even larger pool of resources, skills, and experience. Another side effect resulted from the fact that we inverted the model for staffing projects by giving the individuals the responsibility for indicating interest in a particular project. There was one project that was announced that received very little interest from employees and ended up not being staffed. In this situation, self-selection occurred, because people felt that the project was not compelling enough to be pursued.

We also can imagine a situation where even compelling projects are not staffed, simply because all resources have been assigned to other activities. In that situation, someone needs to decide to reprioritize existing work or put the project on hold until resources become available.

Challenges

The major challenge to this model was in changing the responsibility and decision-making power for what projects employees would work on. Managers and program managers were accustomed to telling employees what they should be working on and making assignments. In the alternative staffing model, the employees could choose what they worked on and in one case even shut down a project because of lack of interest. Managers acted more like a coach rather than a boss.

The fluid nature of the organization, ambiguity of what the end-result might be, and inexperience with the virtual team model was often confusing to employees. Because so much responsibility was placed on the employees for seeking out projects themselves, newly hired employees sometimes felt isolated, since they didn't belong to an intact team that was working on the same project together.

Best Practices

Although it would be difficult to fully implement the alternative staffing model across an organization today because it is so different from the way an organization is traditionally managed, there are many ideas that can be leveraged and used immediately.

Advertise projects – Making projects visible to a larger audience can generate input and interest from people who may not be related directly to the activity. This also has the potential to leverage previous experience to help reduce redundancy, since someone already may have done work in this area.

Catalog skills – Publishing and keeping track of what each individual knows helps people find the right resources quickly. It also helps individuals and their managers create development plans to add or enhance skills. One potential down side to this catalog is in generating too much demand for a subject matter expert's time.

Act as a coach – Balancing project responsibility with coaching responsibilities for a manager ensures that employees are receiving ongoing development and training opportunities. It also evolves the relationship between manager and employee to be less directive and more collaborative.

Hold peer reviews – Even if people are too busy to participate directly in a project, holding periodic reviews with your peers gives them an opportunity to provide input and contribute informally. This is an excellent way of gaining additional perspectives and adding diversity to a project.

Extreme E-Lancing

Can we imagine a world where virtual teams are the primary organizational model? It's unlikely that all organizations will become virtual, but the trend is moving toward a more balanced situation where virtual teams are not just a novel and specialized situation, but rather as common as co-located teams.

One example of a massively virtual workforce is illustrated by a company called Manpower Inc. Its Website describes the company as "a world leader in the staffing industry, providing workforce management services and solutions to customers through 4,000 offices in 63 countries. The firm annually provides employment to two million people worldwide and is an industry leader in employee assessment and training."

Chapter XXV

Kids of the Internet Generation

A few years ago, one of us was at a neighbor's house watching a member of our future workforce in action. She was about 14, sitting in front of her computer, talking on the phone with a couple of her friends through three-way calling, as they browsed a Web site together. Every few seconds, she would respond to one of the five instant messaging chats that she had open on her screen. I thought to myself, "This girl knows how to work in a virtual team."

Technology Still Doesn't Solve All Problems

A couple years later, we hired some new college graduates that had grown up in an environment similar to my neighbor's. They were tech savvy and comfortable with the tools of virtual teams. We hired them into a global distributed team and assumed they would be right at home. What we didn't realize at the time was that we had ourselves believing in one of the myths of virtual teams: technology solves all problems.

We found that the new hires were smart and very comfortable with technology. But they didn't have the network of contacts that the experienced employees had established throughout their career. The technology was familiar to them, but the process of getting work done in a virtual environment was not. They had the cultural background of being in a virtual environment but had never worked in one. They were prepared to work in such an environment, but they were not yet able to do it themselves.

To balance the gap between their technology readiness and ability to work in a virtual team, we gave them a chance to work on projects together and with co-located employees, while at the same time assigning them a mentor to help them navigate the organizational structure and processes of a large company and helping them build their network of key contacts. Sometimes the mentor would be local, and sometimes they would be remote. Both seemed to be equally effective.

Once they understood the process of working in a virtual team, the new hires thrived in the environment and, as expected, were typically the first to try out new collaborative technologies.

Staying in Touch with the Family

My nine-year-old daughter and six-year-old son have relatives throughout the United States. They keep in touch with them through short instant messaging chats every once in a while. My daughter once had a school assignment to interview her grandfather, my dad, in order to understand our family heritage and get a feel for what it was like to be a kid while he was growing up. Although my dad lives only an hour away, and it would have been easy for my daughter to have a face-to-face interview with him, he has almost completely lost his hearing. We decided to take advantage of my daughter's familiarity with instant messaging and hold her interview online. While she is learning how to use technologies that will help her be successful in a virtual team, she is also learning another very important lesson—that family members are the people that you are connected to, not just the ones that you see or live with in your house.

Chapter XXVI

Virtual Meeting Five-Year Vision/Narrative

Virtual meetings work pretty well today, using a combination of audiobridges, Web casting, and videoconferencing. But there are still a lot of inconveniences, minor annoyances, and technical limitations. If we look ahead several years, how might the virtual meeting experience be improved? This chapter looks at a possible virtual meeting scenario in the year 2010, where we look at the experiences of various participants in a virtual meeting. All of the technologies that make the following narrative possible are available to some extent today.

The meeting is a staff meeting with 16 participants:

- Four, including the driver, are in a car
- Six are in a conference room
- Five are telecommuters, all connecting from their homes
- One is on vacation in Death Valley, but doesn't want to miss this important meeting. He is hiking in Ashford Canyon, a particularly isolated part of the National Park.

The participants will be viewing some slides, as well as brainstorming some ideas.

On the Road

Four team members are on their way back via automobile from an industry conference. When it is time for the meeting, the computing center in their car connects to the meeting audiobridge through a voice-over IP connection and automatically brings up the data-sharing application. For the first half-hour of the meeting, the car will be traveling in an urban area with high-bandwidth wireless. On wireless handheld devices, as well as on the pull-down screen in the back seat of the car, the people in the car not only will see the slides being presented during the meeting, but will also have the option of seeing live video of the conference room participants, or any of the telecommuters. No video is transmitted from the vehicle, but vehicle participants have the ability to advance slides, both on their handhelds and on the rear seat pull-down screen. The driver, of course, will have only an audio connection—eyes on the road is the first priority.

After the first half-hour of the meeting, the car moves away from an urban area, and the system automatically switches to a lower-bandwidth wireless WAN connection. Video reception is lost, but the abilities to see, advance slides, and hear are still present.

In a Conference Room

Six members of the team decide to meet in a conference room located at a central site in order to take advantage of the wide-screen, flat-panel HDTVs located in the conference room. Since the conference room is connected to a high-speed LAN and WAN, audio, data, and video are all available, either through a laptop PC, a handheld, or on the wall. As the telecommuters join the meeting, a video of them appears on the screen in a Hollywood-Squares-type arrangement. Since the conference room has multiple flat-panel televisions, each one can display a different image—the slides used during the meeting, the conference room video feed, the video feeds from the telecommuters, or a map of the United States that shows the current GPS position of the participants that have connected from the car, as well as the hiker in Death Valley.

Part of the meeting is dedicated to a presentation by a sales rep that is trying to sell the team's company on an expensive service contract. The team decides that, while the sales rep won't be physically in the conference room, they still would like the ability to "look the rep in the eyes." For that portion of the meeting, the sales rep will connect to the conference room through a 3-D holographic technology that makes it almost seem like the rep is actually present. The technology, still pricey, even in 2010, is used only for situations in which a negotiation is in process, or where there is expected to be a somewhat adversarial relationship between two parties that are not co-located. The telecommuters will see the sales rep through video conferencing, but not as a holographic image. The travelers in the car, and our hiker in Death Valley, will have to do without any visual representation, either 2-D or 3-D.

One of the conference room participants is a visitor to the facility, and she wants to print a copy of the slides during the meeting. From her laptop, she requests a graphical map of all printers on the same floor of the conference room. Upon pressing the print button, the map shows the quickest route from her location to the printer.

From a Home (Telecommuting)

When the meeting is ready to start, each of the home users will click on a "start meeting" button, which will establish the necessary audio, data, and (if possible) video connections. While each of the five telecommuters has broadband, not all have equal bandwidth. In each case, their systems automatically sense the available bandwidth and decide whether this will be an audio/data-only meeting for them, or whether they also will be able to send or receive video images. Like their counterparts in the conference room, they'll be able to switch among the data sharing application, video feeds from the other telecommuters, a map of the location of the car participants, and a video feed from the conference room. Each of the telecommuters has the option to direct the video from the meeting through their home television, watch it on a computer screen, or both. Audio can be played through headphones, a home television, a home computer system, or a home stereo system.

One of the telecommuters is the note-taker for the meeting. One of the modules of the data-sharing application allows the lucky scribe to capture the notes and have other people on the team see them as desired. At the end of the meeting, the notes are automatically placed on the team's internal Web site and will become available once the team manager has approved them.

Each participant in the meeting is automatically represented in the data-sharing application. When they speak, the data-sharing application indicates their names, locations, and photos. Of course, for those that have incoming video, they also can see a live video of the speaker, if the speaker is one of the telecommuters or is in the conference room.

Hiking

Our Death Valley hiker straps on his water pack, handheld computer/GPS system, emergency medical kit, dried fruit pack, and two other items needed by every hiker—a headset and a screen visor insert for his glasses. At the time of the meeting, the handheld automatically searches for a broadband LAN or WAN signal and, finding none, attempts to find a regular cell phone signal. Failing that, it queries the hiker vocally (through the headphone) and visually (through the insert screen in the glasses). "Do you want me to create a satellite connection for $2.50 per minute?" He answers "yes" through the headset microphone and is connected to the meeting. The hiker has both an audio connection (through the headset), and a view of the slides (through the inset visor). It is possible for him to advance the slides through voice commands to their handheld via the headset microphone, but luckily he won't have to today. The hike is up Ashford Canyon, which climbs 1,100 feet in a mile—he doesn't want to have to do too much talking on this hike.

Other participants in the meeting will be able to track our hiker's progress up the canyon through a GPS/topographic map combination. The other 15 people in the meeting think our hiker is incredibly geeky.

Chapter XXVII

Do Virtual Teams Ever Need to Meet Face-to-Face?

In the prologue of this book, we asked the question, "Do virtual teams *ever* need to meet face-to-face?" Certainly, conventional wisdom has it that successful virtual teams need to meet face-to-face from time to time for things like project kickoffs, milestone meetings, and project celebrations—we've captured that viewpoint in earlier chapters of this book. But going forward, is it really a requirement, or as virtual teams become the norm, will the need fade? There is some evidence that in many organizations, meeting face-to-face is no longer a requirement for virtual teams.

A recent study at the University of Nijmegen, The Netherlands, indicates that the face-to-face culture is important in hierarchical companies that evaluate individual (rather than group) achievement, but not important in team-oriented companies with flatter hierarchies. It is often postulated in academic articles (and used to support the idea of periodic face-to-face meetings for virtual teams) that "85% of human communication is nonverbal." The University of Nijmegen study found that while the importance of nonverbal clues is important in situations that require negotiation or where an adversarial situation is likely, they are much less important in a virtual team situation where everyone has a common goal (and knows that evaluation will be based on the team, not individual, output). The conclusion of the University of Nijmegen study follows:

> In conclusion, it is my contention that the one-sided stress in literature on the necessity of rich media seems to indicate the implicit assumption

> that communication is a two-way monologue instead of a dialogue, and trustability is low. After all, transmission of as many as possible cues is most functional in circumstances where competitive relations prevail, and detailed knowledge of other actors is essential. Recognizing the advantages of a change to post-bureaucracy and institutionalized dialogue opens the prospect of less demanding forms of communication and easier implementation of virtual, not co-located, teams. (Smagt, A.G.M. van der (Forthcoming), Enhancing virtual teams; Social relations vs. communication technology. In: Industrial Mgmt & Data Syst)

Right now, we believe that we are in a transition period where many companies are half in the co-located team mindset and half in the virtual team mindset. Thus, not all vestiges of the co-located mindset have yet faded away. Once companies have moved beyond this transition period, we feel that the need for non-revenue generating face-to-face meetings for virtual teams will fade, and perhaps even disappear. This will be accelerated as the expense, inconvenience, and security issues associated with travel continue to proliferate.

Tools Appendix

This appendix will provide more detail on some of the technology tools available to virtual teams. Note that this is not intended to be a specific endorsement of any product, nor is the list designed to include all features that might be available—it is simply a brief survey of the authors' experiences in using the tools within virtual teams.

Audioconferencing bridge lines – Audioconferencing technologies allow virtual meeting participants to gather easily via telephone to conduct meetings. Meeting attendees are provided a dial-in telephone number and a meeting identification number in advance of the meeting. After dialing the phone number, attendees enter the meeting ID via their phone's numeric keypad to gain entry into the meeting.

The capabilities of audioconferencing services vary (as well as their costs); however, most service providers have services that can accommodate anywhere from several attendees to hundreds of attendees. Audioconferencing services can have a variety of features, depending on the service provider in use. Here are some of the features that we've found most useful in conducting virtual meetings:

- Individual line mutes – This feature allows individual participants to completely mute their lines (both talking and line noise) via the service.

(Note that the mute feature on telephones usually only mutes the talking from that line, not noise on the phone line itself.)

- All participants mute feature – Some audioconferencing services have a feature that allows the meeting coordinator to immediately mute all lines at once. This can be particularly helpful if there is excessive line noise on the call that cannot be narrowed down to the individual, someone puts the conference call on hold (and starts music playing in the background), and so forth.
- Breakout sessions – Some services offer breakout sessions that allow smaller groups from the meeting to gather individually to meet.
- Recording – Some services have a feature where the call can be automatically recorded for off-line listening. This is particularly helpful if you know that you have some team members who cannot attend due to time zone issues.
- Toll-free and international numbers – In addition to a main dial-in number, many services can offer toll-free and international numbers, particularly useful if you have mobile users or a virtual team that crosses geographies.
- Roll call – This feature allows meeting coordinators to listen to a recorded listing of all attendees on the line, culled from how attendees identified themselves when they joined the call.

Meeting management software – Meeting management software provides another layer of features for virtual team meetings, enabling functions such as application sharing, shared viewing of slide sets, and so forth. Use of meeting management software typically involves all participants connecting to a central location (either a PC host or Web site) to join the meeting.

Some examples of meeting management software that the authors have used include Microsoft Netmeeting (for smaller group meetings), Hewlett-Packard's Virtual Classroom (larger groups), and WebEx (larger groups). Here are some of the generic features that meeting management software (depending on the product in use) can potentially provide:

- Application sharing – Application sharing allows all meeting attendees to see simultaneously the same application displayed on their screens. Applications often can be shared in a way that allows all attendees to edit the file, if necessary.

- Shared notepad or white board – This feature allows all participants to see a common area for compiling notes, brainstorming, and the like.
- Hands-up feature – This feature allows participants to enter a queue to ask questions during the meeting. A feature like this is particularly useful for queuing questions or comments during large meetings.
- Attendee list – Attendees typically will be able to view a list of all other attendees during the meeting.
- Chat feature – Allows participants to send electronic messages to other attendees during the meeting.
- Message area – Presenters can utilize a message area that all participants can view during the meeting for general announcements. This feature, for example, can be used to advertise how to mute/unmute an individual line during the conference call, post a picture of the presenter, and so forth.

Team spaces – Team spaces is an umbrella term often applied to any number of software solutions and Web-based services that provide a tool-set for virtual teams (or any team, for that matter) to collaborate together off-line. Features often include:

- Shared file spaces – Shared file spaces provide an area for team members to place files for others to view, modify, and so forth, as well as for archival purposes.
- Notepad area – An area where team members can brainstorm ideas off-line.
- Discussion or chat area – A place where team members can initiate off-line conversations on areas of interest to the team.
- Shared Web browsing – Some team spaces will allow a group to browse the Web together, viewing the same Web pages, and so forth.
- Shared calendar – A shared calendar can allow team members to track meetings, project deadlines, and the like, in one place.

Instant messaging – Instant messaging is a software tool that allows users to view what is often labeled as presence information about their colleagues and to exchange short messages. While instant messaging first emerged as a way for Internet users to determine whether friends or family were currently online, it can also provide valuable functionality to geographically distributed teams within corporate environments. Instant messaging can be used to help simulate the type of informal chatting that might take place around the water cooler, in the hallway, and so forth, within co-located teams. Some of the features of the typical instant messaging product include:

- One of the core features of instant messaging is the ability for individuals to quickly exchange short text messages. Team members can use instant messaging to ask quick questions and exchange information, as well as provide a way to allow geographically distributed team members to interact more informally, like co-located teams might chat around the water cooler. Instant messaging can also be very valuable during virtual meetings, allowing presenters or attendees to exchange messages (e.g., "Speak up, we can't hear you," "There are 40 attendees on the line," etc.) behind the scenes.
- Users typically can utilize descriptive designators such as online, off-line, on the phone, away, busy, and so forth, to highlight their current availability within instant messaging products. Remote team members are then able to get a general idea of a colleague's availability similar to the way that co-located team members would glance across the cubicles to see if someone was in their office or on the phone. For example, if a colleague you would like to talk to appears as online, you can send them a quick message (i.e., "Would it be okay if I called you to talk for 15 minutes?"). They would have the option of replying back, perhaps something like, "Sure, give me a call," or "How about we link up in 30 minutes?" and so forth.
- Some instant messaging products allow additional users to be invited to join a chat session. In this way, three or four team members can be involved in the same electronic conversation.
- Some instant messaging clients also provide the ability to exchange files and photos as well as send simple messages to cell phone users.

Glossary of Terms

All-hands meeting – Informational meetings, often held on a quarterly basis, where all members of a particular organization (or, in some cases, the entire company) are invited to attend to hear general organization announcements such as updates on key initiatives, discussion of key organizational goals, general announcements, companywide updates, etc. Often referred to as coffee sessions, coffee talks, or town hall meetings, as well.

Audioconferencing – A service that allows virtual meeting participants to gather easily via the telephone to conduct meetings. (See Tools index of discussion of features and functionality.)

Center-of-the-universe syndrome – Center-of-the-universe syndrome can emerge when there is an area of large concentration of co-location for a particular virtual team. Team members in that location often develop the mindset that their location is the "center of the universe" and that any face-to-face meetings planned should occur there, virtual meetings should be scheduled to best accommodate that time zone, and so forth.

Coffee sessions, coffee talks – See all-hands meeting.

Co-located Team – A team where all team members are located in the same floor, building, campus, etc., and are able to easily gather for face to face meetings.

E-lancing – A virtual team model where freelance workers and employers come together in the open market through the Internet to form virtual working teams. This term was first coined by Professors Tom Malone and Rob Laubacher at the Massachusetts Institute of Technology.

Firewall – Hardware or software solution that is used to prohibited unauthorized access to a network or computing system.

Geographically distributed project teams – A project team comprised of individuals from different organizations (with different management reporting structures) brought together to deliver a specific set of results.

Instant messaging – Software solutions that allow team members to exchange real time messages (chat) and presence information (I'm off-line, online, busy, away, on the phone, etc.).

Manage by objective – Management style that focuses on rewarding those who meet clearly articulated goals and objectives.

Meeting management software – Software or solutions that help with facilitation of virtual team meetings, enabling functions such as application sharing, shared viewing of slide sets, and so forth.

Organizational distributed teams – Teams in which members share a common management reporting structure; however, the members are geographically distributed across multiple sites or locations.

Plan of record (POR) – A list of active projects and programs and their associated completion dates. Typically, each entry in a POR also includes a contact name or the name of the project owner.

Presence - Term applied to the features of instant messaging that allow users to determine whether colleagues are online, busy, on the phone, away, and so forth.

Project launch – The initial meeting for a project team. Agenda items at a project launch typically include tasks such as introducing project team members, outlining project goals, developing project plans and schedules, assessing project resources, and so forth.

Remote manager – An individual who manages employees that do not reside on the same floor or in the same building or campus area, thus limiting opportunities for face-to-face interaction.

Teamware or **teamspaces** – An umbrella term often applied to any number of software solutions and Web-based services that provide a tool-set for virtual teams (or any team, for that matter) to collaborate together off-line. (See Tools index of discussion of features and functionality.)

Telecommuter – An individual who works from their home.

Town hall meeting – See all-hands meeting.

Virtual communities – Groups of individuals that form around a common, shared interest and interact through electronic toolsets (e-mail, chat rooms, etc.). Within corporations, communities can be formed around areas such as shared organizations (all employees on the XYZ team), job functions (sales representatives, lab engineers, etc.), areas of interest (information technology, software development, knowledge management, etc.).

Virtual ecosystem – Groups of organizations that act together primarily through electronic means. The virtual ecosystem can include one or more organizations that act as suppliers, consumers, or partners.

Virtual meetings – Meetings conducted with no face-to-face interaction among the attendees. Virtual meetings are facilitated by tools such as audioconferencing, meeting management software, and the like.

Virtual organization – An organization whose members are geographically distributed among sites and alternate work locations (telecommuting, etc.).

Virtual team – A team with members that are geographically distributed across more than one location. Virtual teams can include geographically dispersed teams, where team members live and work in different locations/states/countries from each other; teams with telecommuters (a form of geographic dispersion in itself); teams formed horizontally across vertical organizations (project teams, task forces, etc.); or teams formed across different companies.

Water cooler experience – Phrase often used to describe the types of information interaction that occurs in breakrooms, coffee areas, water cooler, and so forth, among members of co-located teams.

About the Authors

Robert C. Jones works for the Hewlett Packard IT Enterprise Architecture group, where he manages several internal forums and newsletters on the topic of emerging technologies. He has spent most of his 13-year career in HP IT working in advanced technology groups. During that time, he has worked exclusively in virtual teams (both geographically and organizationally). He has published numerous adult Sunday School courses, which are in use in churches of various denominations throughout the United States, Puerto Rico, Canada, England, Scotland, Ireland, Pakistan, Philippines, Togo, and Kenya. He is also President of the Kennesaw (Georgia) Historical Society, for whom he has written several books, including "The Law Heard 'Round the World - An Examination of the Kennesaw Gun Law and Its Effects on the Community", "Retracing the Route of the General - Following in the Footsteps of the Andrews Raid", and "Kennesaw (Big Shanty) in the 19th Century". He has also written several books on ghost towns in the Southwest, including in Death Valley, Nevada, Arizona, New Mexico, and Mojave National Preserve.

Robert Lee Oyung is currently an R&D Lab Manager at Hewlett-Packard. He has over 15 years of experience in Information Technology. He has held global management, strategy, and engineering roles in mobility, telecommunications, messaging, collaboration, and IT infrastructure services. He is a member of the Advisory Board for the Mobile Enterprise Alliance. He has managed and participated on virtual teams for the past 10 years. Rob is located in Palo Alto, California and holds a Bachelor's degree in Computer Science from the University of California at Berkeley.

Lise Shade Pace currently works in the Enterprise Architecture area within HP IT. She has been a full time telecommuter for seven years, and she has worked on virtual, geographically distributed teams for about 12 years. In her assignments within HP IT, she has focused on examining emerging technologies and their potential impact. She resides in the metro-Atlanta, Georgia area and is a graduate of the Georgia Institute of Technology and Oglethorpe University.

Index

W

Y